T0344370

Let There Be Light

Let There Be Light

Physics, Philosophy & The Dimensional Structure of Consciousness

Stephen J. Hage

Algora Publishing
New York

Library of Congress Cataloging-in-Publication Data —

Hage, Stephen J.
Let there be light: physics, philosophy & the dimensional structure of consciousness / Stephen J. Hage.
pages cm
Includes bibliographical references and index.
ISBN 978-1-62894-030-5 (soft cover: alk. paper) — ISBN 978-1-62894-031-2 (hard cover: alk. paper) — ISBN 978-1-62894-032-9 (ebook)
1. Consciousness. 2. Physics—Philosophy. 3. Quantum theory. I. Title.
QC6.4.C57H34 2013
128'.2—dc23
2013027403

Printed in the United States

This book is dedicated to my beautiful wife Connie, without whose constant support and forbearance I would not have been able to write it.

Acknowledgements

When I read *Transcendence of the Western Mind*, I contacted Samuel Avery via email, in the hope he would he would take the time required to answer some of the many questions I had regarding the work. What struck me, more than anything else, about his answers was how kind he was. His answers always displayed immense patience and thoughtfulness. They always clarified, and expanded my understanding of The Dimensional Structure of Consciousness and helped me to grasp the deep and important principles he laid down in relating how the universe truly works.

As I think back on those many emails I'm constantly reminded of the Buddhist aphorism: When the student is ready, the master appears. I was ready and Samuel Avery appeared.

Later, Sam invited me and several other people who also communicated with him regarding Transcendence of the Western Mind, to work with him on the new book he was writing—Buddha and the Quantum. The experience was exhilarating.

When Sam was on his book tour for Buddha and the Quantum one of his stops was Southern California. I invited him to stay with my wife and me instead of a hotel and he agreed. When I finally got to meet him, I discovered that he truly is as kind, thoughtful and patient as his emails had indicated. I truly believe he is a genius. The master appeared and showed up on my doorstep and I will be forever grateful for his kindness, patience and wisdom. While he visited with me, as we sat together, in my back yard, smoking cigars and sipping fine Kentucky whisky, I asked if he would be willing to write the forward for this book and, of course, he agreed. Thank you Sam.

When I decided to write this book I asked Sam if he would be willing to help me. Being Sam, of course he agreed. He also recommended I ask those who worked with him on Buddha and the Quantum to work with me on this book. That was indeed a piece of sage advice.

The individuals who worked with me, in addition to Sam were, Chuck Lynd, George Fowler, and Toby Johnson. They were generous with their time and their advice and, during the writing process, helped me to clarify my ideas resulting in a work immensely better than it would have been without their patient and generous counsel. Thank you Chuck, George and Toby.

I would like to thank Martin DeMers at Algora Publishing for taking the time to read the manuscript and agreeing to publish this book.

I would also like to thank Andrea Sengstacken, my editor at Algora Publishing for all the help she provided in working through the intricacies associated with properly formatting the manuscript for publication. Her skill as an editor shined through during the editing process and I consider myself fortunate indeed to have had the benefit of her substantial talent and skills. All her diligence and hard work helped make the book easier to read and better than it could have been without her effort on my behalf. Thank you Andrea.

And finally, I would like to thank the light of my life—my wife Connie, for putting up with and supporting me during the writing process. I could not have done it without her. When I was working at Cedars Sinai Medical Center as the director of imaging, whenever I mentioned Connie to my secretary, she would roll her eyes and say, "the woman is a saint." About that, she was right.

Table of Contents

PREFACE

This book is a meditation on and an appreciation of Samuel Avery's *Transcendence of the Western Mind.* My intent is to illuminate some of the concepts and ideas he presented in a style that—if I'm successful—will be relatively easy to understand. That's a tall order given that in *Transcendence of the Western Mind*, Avery presents a new myth useful for understanding why matter, as we have come to appreciate it, doesn't truly exist. For most Westerners, the suggestion that matter doesn't exist is so absurd they aren't even willing to consider it. But, what makes his argument so powerful and convincing is, in presenting us with a new myth he doesn't simply present it de novo. Instead he insists on including the physics associated with all the ideas he presents without resorting to mathematics and mathematical formulas which make most people's eyes glaze over the moment they see them. In that respect, I have followed Avery's lead.

For Westerners, the mindset that governs understanding of how the universe works is tied directly to the ideas and myths created by Rene Descartes. He believed that matter exists and helped shape our ideas regarding consciousness with his famous declaration *cogito ergo sum*—I think, therefore I am. We picked up that ball and ran with it which explains why we believe the cup we see sitting on the table is really out there because, after all, we can see it and touch it and, if we wish, do things with it like drink coffee.

Our Cartesian imprinting is very strong, so strong that most scientists believe that matter truly exists. But, interestingly, especially for scientists and particularly for physicists, to their dismay they have discovered it is impossible to scientifically prove that matter exists.

This raises a host of interesting and difficult questions like, if scientists cannot prove that matter exists, why is it that I have no trouble seeing and manipulating objects I assume are composed of matter? For most people this isn't a big deal because they assume matter exists and that it's just crazy to deny it exists.

But physicists aren't most people. Their approach to exploring "what is" is highly disciplined because it relies on the scientific method. And, when they intensify their focus on matter in an attempt to determine exactly what it's composed of, they discover that it disappears.

And, when they manipulate the smallest constituents of matter like subatomic particles, the results of some of their experiments are so counterintuitive and enigmatic they appear mystical.

Some of their experiments actually prove that the presence of an observer changes the outcome, which indicates strongly that our belief in the existence of matter is somehow inextricably entangled with consciousness. For most scientists, and especially for physicists, this situation is highly problematic because physicists restrict their explorations to objective phenomena. Nothing about consciousness is objective so when their experiments demonstrate that consciousness truly is entangled with the results they get when performing those experiments they usually assume one of two positions. If pressed to explain why and how consciousness is entangled with their experimental results the classic physicist to physicist angry reply is "shut up and calculate." If pressed by a non-physicist with the same question they deflect the question or become apologetic because, since consciousness is not objective, they cannot explain why it appears to be deeply entangled with their experimental results.

For non-physicists, the retort "shut up and calculate" means "*don't get hung up on weird and unexplainable experimental results. Instead, focus your attention on the mathematics associated with your experiments and let the equations speak for themselves.*"

In both instances, they are dodging the question.

Samuel Avery refuses to dodge the question. He faces it head-on by including both consciousness and the physics associated with his explanations in a way which supports his new myth and provides a foundational framework for an innovative paradigm he calls The Dimensional Structure of Consciousness. This new paradigm explains not only why consciousness is entangled with the experimental results physicists get when performing certain highly controlled experiments but also how.

His explanations are powerful and satisfying but they require a willingness, on the part of the reader, to abandon his Cartesian imprinting. And while it would be misleading to suggest that the Dimensional Structure of Consciousness proves that matter does not exist, it does provide an extremely convincing argument for the non-existence of matter.

We use myth to explain what cannot be explained in any other way. The existence of matter is a myth but, when old myths fail, they must be replaced by a new myth which better explains our experimental results. The Dimensional Structure of Consciousness is that new myth.

If you decide to come along for the ride, your reward will be a deeper and more comprehensive understanding of how the universe works; a peek under the veil, if you will, which exposes what's going on when hardnosed physicists get experimental results that are enigmatic and mystical.

Foreword

A paradigm shift changes not what you see but from whence you see it. The world looks the same, before and after; what changes is the looker. If you step outside and watch the stars and planets arcing across the sky on a clear night, they rise over the eastern horizon as they did in the time before Copernicus. You see them exactly as Ptolemy saw them. The difference since Copernicus is in who you are and where you stand in the universe.

This is the great burden of a paradigm shift: It explains the world in a new way, but it has also to explain why the old way worked so well for so long. It has to explain what we were seeing before. Copernicus had to tell us why we thought the sun orbited the Earth. The paradigm shift that Stephen Hage and I describe in this and other books on the meaning of modern physics is no exception. The dimensional structure of consciousness that we propose explains everything in the physical world without material substance, but carries the burden of explaining why the world *looks* like it is material. It explains why you touch the flowers on the table where you see and smell them, and why you see the squirrel on the lawn when you hear it, but also why objects in the world *seem* to be there when no one is looking. Because the world of tables and chairs without matter looks exactly as it does with it, we cannot use it to point out the new paradigm. We have to point to scientific observations at the fringes of physical reality that do not fit into the material worldview, exotic phenomena that exist far beyond the familiar world of everyday life. The Uncertainty Principle, the space, time, and mass dilations of relativity theory, black holes, the wave-particle duality, and the non-localities of quantum mechanics all tell us that something is very wrong with our understanding at the extremes of the dimensional world. The essence of our argument is that this is the same dimensional world that we experience in everyday life, and that there is something very wrong—or at least very old fashioned—in how we understand what we experience sitting in the living room or walking down the street. Understood anew, the enigmas of modern physics, like the retrograde motions of outer planets, revolutionize not what is seen, but the mind that sees them.

At the heart of the revolution is a new understanding of dimensions. In the material worldview, space and time define an external universe populated with material objects that exist with or without conscious experience. We question this assumption because a) it is an assumption; b) all of the enigmas of modern physics involve limitations and distortions in space, time, and mass dimensions, and c) the same enigmas involve observation (conscious experience) of physical phenomena. We propose that dimensions, rather than structures of an independently existing universe, are structures internal to consciousness. They are in us, not we in them. More specifically, a dimension is a *potential for sensory information*. This is the key to understanding modern physics. A dimension is not something "out there" that we are in; it is something that structures, organizes, and gives meaning to what is seen, heard, touched, tasted, and smelled. Information is always an actual within a potential: it is a dot that could be a dot or a dash, a yes that could be a yes or a no, a red that could be a green, or a number that could be some other number. Sensory information is an actual point or locus of points within a potential experienced in the form of a dimension. Mass, rather than a measure of material content, is the second time dimension of acceleration. What we experience as the physical world is five realms of sensory experience intercoordinated into five space, time, and mass dimensions. It is the intercoordination that produces the appearance of material substance. Where and when you see flowers on the table is where and when you touch or smell them. Where and when you hear a squirrel on the lawn is where and when you see it. Knowing you will feel an object where you see it gives it the appearance of it being "out there" when you do not see it. (It is only *potential* perception that is dimensionally coordinated: where an object is heard has to do with where it is seen, not what it looks like.) This "dimensional structure of consciousness" is difficult to grasp initially, but far simpler than the material worldview.

The concept of matter is so deeply engrained that we have forgotten that it is an assumption. Matter is assumed to exist outside of experience. It will always remain an assumption because experience is limited to what we actually see, hear, taste, touch, and smell, and no one has ever, or will ever, experience anything beyond experience. There is no way to prove it. The new paradigm does not require this assumption; therefore it is simpler.

Because the everyday world looks the same with either paradigm, the new paradigm cannot be derived from everyday experience. It is superior to the material worldview only because it is consistent with everyday life *and* with the enigmas. It applies to the middle latitudes of space, time, and mass, where we have our being, as well as to dimensional extremes (extreme velocity, size, mass, and distance) where the enigmas occur. Light, for instance, is the visual realm of consciousness. As its particle nature begins to dominate its wave nature it becomes reducible to the minute tactile sensations of individual photons. The realm of visual information is reducible, therefore, to the realm of tactile information at extremely small dimensional values. If the new paradigm is right, the potentials for visual and tactile perception should blend together at this level, which is exactly what happens. Space and time become indistinguishable at precisely the point where visual and tactile information become indistinguishable. The same fabric of space–time that is rigid, absolute, and unchanging on the macroscopic level unravels to nothing below the quantum level. In seeing how the world falls apart at the quantum level we begin to see how it is put together on the macroscopic level.

If the new paradigm shifts the way we think it will—if the world is put together the way we think it is—we are not who we thought we were and there is an entirely new meaning to life. It could not happen at a better time.

But the dimensional structure of consciousness is only Hage's starting point for this book. He takes the ball and runs to places I never thought to go. This is the sign of a good paradigm—a new myth that cuts through cobwebs of outdated thought and opens new continents for human exploration.

Samuel Avery
Hart County, Kentucky

WHAT IS LIGHT?

Most people think they know what light it is. Many think they know what it does. In truth, very few understand it.

Light exists in our religions and our myths. It has caught the attention of physicists and writers of every stripe. It is in our poetry and prose and, the light we can see is in our eyes.

From *Transcendence of the Western Mind* by Samuel Avery, of light we learn this—"The universe unfolds in the touch of light."

From the King James Bible in Genesis, of light we learn this.

1:1 In the beginning God created the heaven and the earth.

1:2 And the earth was without form, and void; and darkness was upon the face of the deep. And the Spirit of God moved upon the face of the waters.

1:3 And God said, Let there be light: and there was light.

1:4 And God saw the light, that it was good: and God divided the light from the darkness.

Ancient Hindu schools of Samkhya and Vaisheshika from around the 6th–5th century BC believed light is one of the five fundamental "subtle elements (tanmatara) out of which emerge the gross elements."[1]

In 55 BC Lucretius wrote: "The light & heat of the sun; these are composed of minute atoms which, when they are shoved off, lose no time in shooting right across the interspace of air in the direction imparted by the shove."[2]

I find it interesting that Lucretius was right about light losing no time. Modern physics has shown that photons are timeless because the moment they come into existence, since they are light, they're traveling at the speed of light. If we could travel at the speed of light, we wouldn't age; we would be timeless, too. When light from a galaxy thousands of light years away reaches your eye, for that light, no time has passed since it left the galaxy from whence it came. Its "now" is the same as it was the moment it left, thousands of years ago. When you see it, its now becomes yours.

We need light to see and to survive. Light from the sun, in addition to allowing us to see, provides energy and warmth. Plants and animals need it to grow and thrive. Life needs light.

When we try to determine what it is, we're faced with enigmas. Physicists have been studying it for centuries and yet its behavior is still, in the 21st century, enigmatic and paradoxical. We still don't understand what light is or why it behaves the way it does.

We know, from physics experiments, that light travels at about 186,000 miles per second. In the late 19th and early 20th centuries, Albert A. Michelson conducted experiments on the speed of light and, in 1926, determined it to be 299,796,000 meters per second.[3]

Physics is the hardest of the hard sciences. At the time Michelson was running his experiments, physicists had a solid understanding of the mechanics of sound waves. They had conducted earlier experiments which showed that sound waves behaved much like water waves. If you toss a stone into a still pond, you will see waves propagate from where the stone entered the water that look like concentric circles. The water is the medium that carries the waves made by the stone.

For sound, the medium is usually air, but sound also travels through other media like glass, wood, steel and water. Physicists were able to determine the speed of sound through air and other media without too much trouble. So, when they began to study the speed of light, they assumed there must be some medium through which light traveled. They further assumed this medium filled the entire universe, since they were able to see stars. They even gave this medium a name; the *luminiferous ether*. That assumption was grounded in logic. With the stone in the pond, the water is what's doing the waving. With sound it's usually air molecules; so, naturally, with light there had to be some medium that waved in the same way that water and air molecules wave.

As physicists continued to do experiments to measure light's velocity through the luminiferous ether, it occurred to them that since it filled all space, everything in space must be moving through the luminiferous ether too, including earth. And, since the earth moves pretty fast around the sun, if they measured the speed of light at six month intervals, they should be able to detect a slight difference in the speed of light when the earth was moving through the ether in the opposite direction. The logic was unassailable but it didn't work. No matter how they set up their experiments or how careful they were with their instrumentation and measurements they could detect no difference in the speed of light.

Clearly something was wrong. Their math was good enough to tell them that, with the instruments they had, they should be able to detect and measure a difference in the speed of light when the earth was moving in the opposite direction through the ether. But, try as they might, they could find no difference. This was a problem. Something had to be waving and that something had to be the ether. So, either the ether wasn't waving or there wasn't any ether. Neither answer was satisfactory even though their careful experiments told them it was so; big problem!

Physicists knew that light behaved as if it were composed of waves because around 1800, Thomas Young performed the classic double slit experiment. He shined a light source through a plate that had two thin vertical slits cut into it. Opposite that plate he placed a screen and what he saw was an interference pattern of dark and bright vertical bars which proved that light was composed of waves that interfered with each other as they passed through the slits. The dark bars were areas where the waves cancelled each other out. The bright bars showed where the waves reinforced each other. Thinkers like Lucretius and Newton thought light was composed of discrete particles but Young's double slit experiment seemed, at the time, to have settled that question. It proved light behaved as if it were composed of waves.

It wasn't until the early twentieth century that Albert Einstein proved that light was composed of discreet particles called photons. Now physicists were faced with another dilemma associated with light. The experiment that Young performed proved light was composed of waves. Einstein's experiment proved it was composed of particles. They were both right. The difference was in the way they chose to set up their experiments. That realization led Einstein to his next intellectual tsunami—Special Relativity.

Does it really move?

Does light really move? Superficially the question seems silly. Of course light moves. If you turn a flashlight on and point it at the wall across the room you will see a splotch of light on the wall. And if you move the flashlight around the splotch will move too. End of story. But, that's what we thought when Young did his experiment which proved light was composed of waves. Then Einstein proved light was composed of discrete particles called photons. Suddenly the story became harder to understand. If you set up your experiment to prove light was composed of particles then it behaved as if it truly was composed of particles and, if you set it up to prove it was composed of waves then it behaved as if it was composed of waves. Both experiments work yet they both prove different things about the nature of light.

Einstein was awarded the Nobel Prize for his work which proved that light is composed of particles. Beyond that, Einstein discovered other properties of light even more bizarre and puzzling than its wave particle duality.

Like most physicists during the early 20th century Einstein was puzzled by the failure of every experiment to discover a medium for light. Common sense said there had to be a luminiferous ether because, without it, light couldn't travel anywhere for the same reason that, absent an appropriate medium, sound can't travel.

I can still remember sitting in high school science class and watching as the teacher placed a mechanical alarm clock under a bell jar connected to a vacuum pump. He had set the clock to go off in ten minutes and we all watched as the pump droned and evacuated the air in the bell jar. Then he shut the pump off and, in a few minutes, the alarm went off. We could see the metal clappers beating furiously against the two large bells on the outside of the clock but we couldn't hear a thing because there was no air in the bell jar; no medium-no sound. This was physics in the grand tradition. When he opened the valve allowing air to rush back into the jar we heard the alarm clock loud and true.

This experiment has been done thousands of times and is a splendid example of how science works. Using the scientific method you formulate a hypothesis, generate a theory and, if possible, conduct experiments to prove or disprove the theory. The hypothesis for sound is there must be some medium to carry it. The theory states, that medium is air. And the experiment with the alarm clock and the bell jar proves the theory. It's all very nice, tight and clean. And the beauty part of the scientific method is anyone can do the same experiment any time and get exactly the same results. That's what makes science so important and powerful. It's also what drives scientists so crazy when it comes to light.

The most powerful and reliable tool scientists have at their disposal is the scientific method. They've used it to study the heavens and build atomic bombs but, when they crank out the good ol' scientific method to study light—light refuses to behave or cooperate. What is it about light?

After thinking about all the experiments that had been done to discover the luminiferous ether Einstein cut the Gordian knot. He said the reason no one can find the luminiferous ether is because there is no luminiferous ether. Light doesn't need a medium through which to travel. But he didn't stop there. He said its speed never changes and established the universal constant c which every good physicist knows stands for the speed of light.

At first blush you may be thinking what's the big deal? The speed of light is constant. So what?

Here's the deal. Imagine you're riding in a car going down the road at sixty miles per hour. You open your window and point a gun down the road in the direction you're traveling. Then you fire the gun and the bullet leaves the muzzle of the gun at 1000 miles per hour. Here's the question. How fast is the bullet traveling relative to the ground? The answer is intuitively obvious. Relative to the ground, which isn't moving, the speed of the bullet is its muzzle velocity, 1000 miles per hour plus the velocity of the car which is sixty miles per hour for a total of 1060 miles per hour. Duh!

Now let's do the thought experiment again. Same situation only this time the driver turns on his headlights. Here's the question. How fast is the light from his headlights traveling relative to the ground? Common sense says the result should be the same as for the bullet. Light travels at 186,000 miles per second or 669,600,000 miles per hour so its speed should be 669,600,060 miles per hour. And while that has to be the right answer for the same reason it is for the bullet; it isn't the right answer. The right answer is 669,600,000 miles per hour or 186,000 miles per second. That's what Einstein meant when he said the speed of light is constant. It wouldn't matter how fast the car was traveling when the driver turned on his headlights. The light from those headlights would be traveling at 186,000 miles per second.

But wait, there's more. What if the car were traveling in reverse? With the bullet we'd have to subtract the speed of the car from the muzzle velocity of the bullet to determine its speed relative to the ground. But, with the headlights, it's the same deal as when the car was moving forward. The speed of light leaving the headlights would be 186,000 miles per second regardless of how fast the car was moving.

Under normal circumstances we have a pretty good handle on how things move. By things I mean tables and chairs, bullets and beans, baseballs, hot dogs, apple pie and Chevrolets. We don't think too much about all this because scientists from Isaac Newton's time through today have conducted millions of experiments with these kinds of things and non scientist types have eaten them, driven them, crashed them into each other and other things like buildings and blown them up. We know about how things move when we interact with them or when they interact with us. We know that, if we see a big truck barreling down the road in our direction, we'd best get out of the way.

But light is different. If it moves, it always does so at the same speed relative to everything. How come that is? Why doesn't it behave like a bullet or a baseball?

As far as things other than light are concerned, for a long time, we were pretty sure we understood everything there was to know about them until Einstein came along with his theory of special relativity. After that, all bets were off.

Before Einstein, physicists believed they'd pretty much figured out all there was to know about the physical universe. Sure, there were a few niggling issues like light and the luminiferous ether but, other than that, they felt they'd nailed everything down. They were worried that there wasn't much left for physicists to do and that, in the future, their primary concerns would be with refining what they already knew down to ever more decimal places with better instrumentation and equations.

Einstein saw things differently.

The Weirding Ways of Special Relativity

Einstein was fond of using thought experiments to reach conclusions. What I said earlier about traveling in a car and firing a gun out the window is an example of a thought experiment. As he continued to think about light and his realization that its speed is constant something else occurred to him. If he was right about the speed of light being constant, then it seemed as if space-time instead of being something that's the same for everybody is more like an invisible bubble that everybody carries around with them all the time. Everything is fine as long as we don't move too fast in relation to each other. Our personal bubbles look the same and it appears we're all in the same bubble.

If, on the other hand, we begin to move very fast weird things begin to happen and the idea that there's only one space-time bubble bursts and falls apart.

Here's what I mean. Let's do another thought experiment. Imagine that you and I are floating in outer space. We're each enclosed in a large Plexiglas sphere about ten feet in diameter. Inside the sphere we have a large clock, a desk, a yardstick, and a chair. Now since this is a thought experiment we can do anything we like. Let's assume we can make our spheres accelerate at will and that we're about 200,000 miles apart and our spheres are oriented so that when we pass we'll be facing each other. Got that? We each have a button which, when we press it, will make our sphere accelerate instantaneously to half the speed of light. We push our buttons at about the same time and begin accelerating toward each other. Here's where things start to get weird.

As we pass each other, I notice that things aren't quite right in your sphere. For one thing, your clock is running slow. For another, your desk is noticeably shorter than mine and so is your yardstick, as long as you have it pointed in the direction you're moving. If you rotate it 90 degrees, it lengthens back to 36 inches but it gets narrower in the direction you're moving. And finally, I can tell that everything in your sphere has gotten heavier. And while that doesn't seem right, it is what's happening. In my sphere everything looks fine, but there are weird things going on in yours.

For you, everything seems fine too. You don't notice that your desk is shorter or that your yardstick gets shorter when you point it in the direction you're moving, and you don't notice that your clock is running slower either. Nothing seems to have gotten any heavier in your sphere. But, when you look at my sphere you notice the same things I saw in yours. My desk and yardstick are shorter. My clock is running slower and everything in my sphere has gotten heavier.

In this thought experiment it's important to understand that what you and I are observing about the other person's sphere is real. Our clocks really are running slower. Our desks and yardsticks really are shorter and everything in our spheres really has gotten heavier. The reason is because we are moving very fast. As we approach the speed of light our personal space-time bubbles get out of synch. And if we continue to accelerate, they stay out of synch. If one of us accelerates even more, coming closer to the speed of light, our personal space-time bubbles move even further out of synch. If my sphere is the one that accelerates, then my clock runs even slower than yours, my desk and yardstick get even shorter, and everything gets heavier. None of this is illusory. It really happens.

When we look at what's happening in our own sphere, because everything seems fine and there are no odd phenomena occurring, we each think the other person is nuts. The problem is we're both wrong. You may think that I'm moving and you're at rest, and I may think I'm the one who's moving and you're at rest. As it turns out, that's not important because there is no such thing as absolute space. All velocities are relative because there is no such thing as a special or preferred frame of reference for the entire universe. It is the reference frame's velocity—*in this thought experiment, the spheres we're in are our reference frames and those frames are in our own personal space bubbles*—that causes physical phenomena to vary like with our clocks, desks and yardsticks; but there is no one box for everybody. Everybody's box or space bubble is different but physically equivalent in terms of velocity.

Let's expand this thought experiment just a bit. There are some caveats associated with equivalence in special relativity. When we looked at each other's clocks as we passed, we both thought the other person's clock was running slow. What about total elapsed time? What would happen if, say, I slowed down to your frame of reference so we could compare notes? Interestingly total elapsed time will be different because we saw each other's clocks running slow so one of us will have experienced more total time than the other but the question then becomes, which one? How can it be possible for an absolute difference in time to exist *in the time we experienced* if motion is purely relative?

And the answer is, wait for it, not *all* motion is relative. Uniform, steady motion in a straight line or constant velocity is relative. But, *here's the important part*, acceleration is NOT relative. We tend to think of acceleration as starting from a standstill and increasing velocity until we reach some predetermined value like going from 0 to 60 miles per hour. But acceleration is much more than just that. It includes going over bumps and around curves and accelerating past 60 to 80 miles per hour and slowing down from 80 to 40 miles per hour. All these events are forms of acceleration and acceleration is not relative.

Positions in space are purely relative. So is constant velocity. But acceleration is *absolute.* The person who accelerates is the one who gets to experience less total passage of time relative to one who does not accelerate.

If I stayed on the ground while you got in your sphere and took off and your trip lasted a week, when you returned you'd tell me you had a pleasant trip but that you didn't want to stay away for more than a week. And I would say, buddy you've been gone for a month. Another very strange thing about all this is, if we were the same age, you'd be younger than me when you returned. As counterintuitive as all this seems, it's important to keep in mind that it's all true. We know this because we've done experiments which prove it to be true.

At speed, the world we're used to becomes like the world Alice encountered when she walked through the looking glass. Things are not what they seem.

When we experience acceleration of any kind, we can feel it. We experience a kinesthetic sensation that tells us we've either sped up, slowed down, gone around a curve or over a bump. Science is unable to explain why this is so. You could say that when we accelerate our cells become compacted but, the question then becomes, how do the cells know? As it turns out there's absolutely no way to explain the absolute nature of acceleration in the material world view.

When you accelerate, what you actually *feel* is a g force and this g force is the *second* time component of your motion the second *per second* as in going from 0 to 60 in 4 seconds. And, the magnitude of what you feel is in direct proportion to your acceleration. If you went from 0 to 60 in 2 seconds instead of 4 you'd feel it even more. As soon as you level off to a constant velocity, you no longer feel it even if it's half the speed of light. What you actually feel when you accelerate is space—time—time. This is motion relative to the entire universe. When your personal space bubble falls out of step with mine what you experience is directly related to what you experience in the tactile realm of perception.

As you think about all this it's important to keep in mind that space, time and mass distortions become *noticeable* at extreme velocities; velocities approaching light speed. But, that's not where they *exist.* They exist at *all* velocities. People passing each other in cars, or even when walking down the sidewalk have slightly different experiences of space, time and mass. They're not noticeable because they're so small and they're so small they can't be measured but they're never exactly the same. What we see all around us as we go about our daily lives, while we're moving, is never more than an extremely good approximation. Space, time and mass are always skewed slightly out of synch.

Our normal, everyday, low velocity experience is a special case that actually exists in a much larger context. We consider it *normal* because it's what we're most familiar with. But, it's important to keep in mind and remember that experiences at half the speed of light are every bit as real.

In his special theory of relatively Einstein gave us something else; his famous equation $E=mc^2$ where E is energy, m is mass and c is the speed of light. This equation makes several astounding statements about the relationship between mass, energy and the speed of light.

- The amount of energy available in the entire universe is constant
- Mass can be converted into energy
- Energy can be converted into mass
- Mass and energy are essentially the same thing
- The relationship between mass and energy is related to the speed of light
- Mass and energy can neither be created nor destroyed—they are constant

That's a lot of information packed into one little formula. One familiar proof of $E=mc^2$ is the atomic bomb. When an atomic bomb is detonated, the mass of plutonium in its core is instantaneously converted into energy resulting in a blast strong enough to destroy a city. That core is small enough for you to be able to hold it in your hands although, if you did, you'd die of radiation poisoning.

As astounding as all this is, $E=mc^2$ is much more than a recipe for building atomic bombs and reactors. It's a statement about the deep relationship between tactile and visual consciousness. In a very elemental way it speaks to what we are and what we are becoming. The extent to which we ultimately understand it will determine how we live or die.

"The equivalence of mass and energy is a question of being that we have answered only in terms of doing."[4]

Light & space time

As we go about our lives, most of us give very little, if any, thought to subjects like light, space and time. For most of us, these phenomena just are. But, like light, space and time become troublesome when we begin to examine them and ask questions about what they are and how they work.

Time is particularly problematic. It is integral to many equations associated with physics, classical mechanics and quantum mechanics. It is important for our understanding of what the universe is and how it works. But, given all that, I find it interesting and troublesome that, unlike sound or heat or light, time cannot be detected. Humans and other animals have organs that can detect sound, heat and light. Scientists have built instruments that can detect sound, heat and light. But, no one has ever built an instrument that can detect time; and we have no organs that allow us to detect it either.

What about a clock you ask? Sorry, no cigar. A clock or wristwatch is nothing more than a small model of the earth rotating on its axis. For most clocks, one revolution of the earth is equivalent to two trips of the hour hand around the face of the clock. It takes 24 hours for that to happen. Clocks and wristwatches don't detect or measure anything. They simply represent the earth spinning on its axis.

I like what St. Augustine had to say about time. "What then is time? If no one asks me, I know: if I wish to explain it to one that asketh, I know not."[5]

It's the same today. Theories and ruminations abound but it seems clear, to me anyway, that no one really knows what time is.

One of my favorite quotes about time is: "Time is nature's way of keeping everything from happening at once." It's been attributed to Einstein, John Archibald Wheeler and Woody Allen. It's pithy, funny and it makes sense but it certainly doesn't explain what time is.

Space is similarly troublesome. If we don't ask questions about it we know what space is. When we try to explain it scientifically we run into problems.

Einstein combined space and time into a single entity, a sort of fabric, known as space–time. It's important to understand, however, that the actual existence of space–time is still speculative.

When we study light we get the impression that it moves through and exists in space–time; but more on that later.

Now!

Time appears to flow, and more than just that, it appears to flow in a recognizable direction; from present to future. Physicists refer to that flow as the arrow of time. Examples of the arrow of time and how it's entangled with entropy abound. Drop an egg on the floor and, you've got a mess. You also know there's no reasonable chance that egg will reassemble itself and fly back into your hand. You're going to have to clean up the mess. It's the same deal with spilt milk and smoked cigarettes.

If you put energy into a physical system like, say, winding a mechanical clock, the system becomes highly ordered. But given enough time, it will descend into a more chaotic unwound state. Its entropy will increase and, until or unless someone winds it up again it will remain entropic. Same deal with putting a battery into a wristwatch or camera. Physical systems of every stripe that are highly ordered descend into chaos. That's what the second law of thermodynamics is all about.

We find ourselves embedded in all this. It's part of being alive and we learn to deal with it. We're constantly presented with examples of the arrow of time and entropy and for most of us it's okay. We don't question what's going on. We simply accept what we see and move on. Most of us just clean our houses or apartments occasionally, or pay someone to do it for us. There's no way around entropy.

For scientists, things aren't so easy. If a physicist drops an egg, he'll be confronted with the same mess you or I would be confronted with. The problem for physicists is that while time is a key element in many of the experiments they perform and the equations associated with those experiments, when they examine the equations carefully their calculations tell them that the results are time invariant. In other words, the equations work just as well with time moving backwards as they do with time moving forward. Something's wrong with that because, after all, broken eggs and spilt milk are phenomena everyone is familiar with but, if we use math to model what happens when we break an egg or spill milk, the equations work just as well backwards as forward. We get a feel for how strange this is if we watch a movie of a dropped egg or a glass of spilt milk, run backward. In the reverse movie the egg does reassemble itself as does the shattered glass and the milk. The movie behaves like the equations in that it works just as well in reverse as it does running forward.

If asked what the word now means, most people would say they understand the concept. Now means now. But, like time, when you begin to examine the idea of now carefully you're faced with the same kind of problem St. Augustine had when asked about time.

Now is all we ever have. There is no past and there is no future. There is only now. We are trapped in now. The past is something you remember now. The future is something you think of or imagine now.

Your first kiss is something you can recall now but it's gone-forever. Winning the lotto is just one of an infinite number of possibilities that may happen in the future. For all of us, from the moment we're conceived, now is all we ever truly have. The question then becomes: How come that is? Why are we always trapped in now? Is there a physical explanation?

I believe there is.

> "Einstein proclaimed that all objects in the universe are always traveling through space-time at one fixed speed—that of light. This is a strange idea; we are used to the notion that objects travel at speeds considerably less than the speed of light. We have repeatedly emphasized this as the reason relativistic effects are so unfamiliar in the everyday world. All of this is true. We are presently talking about an object's combined speed through all four dimensions—three space and one time—and it is the object's speed in this generalized sense that is equal to that of light."[6]

If that statement by Brian Greene, from *The Elegant Universe*, is true, and I believe it is, then it's entirely possible we are trapped in now for the same reason photons are trapped in now. We're moving through space-time at the same speed they are apparently moving through only space. The key difference is, photons always and only move (or appear to move) through space so they never get the chance to move through time which is why they are timeless. They can do that because they have no mass.

Because we and other objects have mass we are disallowed from moving through space at the speed of light. So to the extent any object is moving through space at less than the speed of light the remainder of that object's speed, which brings it to the speed of light, is moving through time.

This is a very strange idea but, it becomes a bit easier to wrap your brain around it if you remember there is no such thing as stillness. On the macroscopic scale of existence where we live we are sometimes confronted with what looks like stillness. But the truth is everything in the universe is always and only in exquisite motion. The earth is constantly rotating on its axis as it whips around the sun. Those two motions alone represent a significant amount of zip and spin. But wait, there's more. The entire galaxy is rotating and moving through space as space constantly expands.

If we look at a block of steel sitting on a table it appears quiescent. But, in addition to moving through space for the same reasons we are, if we examine its atoms we discover that the electrons are orbiting the nuclei and that the nuclei are furiously exchanging particles.

There is no stillness on the planetary, solar system, galactic or quantum levels. I believe it's important to remember and think about this any time you're confronted with a situation that appears to represent stillness. Stillness does not exist.

Myth

Much of what follows is based on my interpretation of Samuel Avery's book *Transcendence of the Western Mind*. It has had a profound impact on how I see the world and my understanding of how the universe works.

There was a time when we believed the earth was flat. It was a myth but, absent any other interpretation life went on and everything seemed to work out fine. The earth's flatness seemed intuitively obvious as long as we didn't move too far in any one direction for too long.

There was also a time when we believed the earth was the center of the universe and that all the heavenly bodies, including our sun, revolved around the earth. The myth was supported by being able to watch the sun rise in the east and set in the west every day. The belief was so strong it became part of religious dogma. It was dangerous to challenge that dogma, so much so, that when Copernicus came to believe the earth orbited the sun rather than the other way around, he held off on publishing his thoughts about the matter until just before his death. After his ideas were published, Galileo was arrested and spent the rest of his life under house arrest for believing that the earth wasn't the center of the universe. Myths die hard.

We use myth to explain what cannot be explained in any other way.

> "Orderly and meaningful existence is impossible without myth. Myth organizes life into recognizable patterns and creates within itself a structural basis for belief and action, the means by which we stay sane and alive. Myth is understanding; it connects doing with being in everyday life.
>
> If a myth begins to fail, a new myth will rise in its place. But the new myth must be better than the old one...The new myth will be far more than science and far more than anything we can say here. But it will begin with science and physics in particular, because physics is the connection between

> the mental and the material, the interface between self and world, and thus the link between two realms that are no longer separate. The great change will happen here. A new mythology of life will begin with physics because physics is the study of what is. Physics is our approach to creation. No one will be more surprised than the physicists."[7]

The existence of matter is a powerful and persistent myth. Most people assume matter exists "out there" in the universe. You see a cup sitting on a table. If you reach your hand out toward the cup you can feel it. If you ask a friend whether she sees the cup she will tell you she does. If the cup is filled with coffee, when you pick it up and bring it to your nose, you will smell the coffee. If it is hot you will feel the warmth of the coffee and, if you take a sip you will taste the coffee's bitterness. When you put the cup back down on the table you will hear a sound when it touches the tabletop. Most people are also relatively certain that if they leave the room where the cup is sitting on the table, the cup will still be there after they've left.

As long as we don't question or examine them too carefully, our ideas about the existence of matter and the myths that support them work pretty well; well enough to get us through the day without too much trouble. But, if we begin to ask questions about the existence of matter we run into problems.

We think we know what matter is because we can manipulate it. We can shape it into objects like cups, computers and atomic bombs and we can use those objects to do things. And even though most people would think that questioning the existence of matter is insane, the reality is there's no way to prove it exists other than perceiving it using one or more of our five senses.

In the 18th century Bishop George Berkeley questioned the existence of matter as something that exists "out there" independent of anyone perceiving it.

> "...I see this cherry, I feel it, I taste it: and I am sure nothing cannot be seen, or felt or tasted: it is therefore real. Take away the sensations of softness, moisture, redness, tartness and you take away the cherry. Since it is not a being distinct from sensation; a cherry, I say, is nothing but a congeries of sensible impressions, or ideas perceived by various senses; which ideas are united into one thing (or have one name given them) by the mind; because they are observed to attend each other. Thus when the palate is affected with such a particular taste, the sight is affected with a red colour, the touch with roundness, softness, &c. Hence, when I see, and feel, and taste, in sundry certain manners, I am sure the cherry exists, or is real; its reality being in my opinion nothing abstracted from those sensations. But if by the word cherry you mean an unknown nature distinct from all those sensible qualities, and by its existence something distinct from its being perceived; then indeed I own, neither you, nor I, nor any one else can be sure it exists."[8]

What Berkeley is saying is the existence of the cherry *is* our perception of it. It exists *because* we perceive it using one or more of our five senses. The touching, tasting, feeling, smelling and hearing *are* the cherry; and absent any sensory perception of it, the cherry doesn't exist.

Very few people are willing to accept Berkeley's opinion regarding the existence of the cherry. Fewer still are willing to accept the extension of his opinion to include all objects that we perceive. And yet, it's impossible to scientifically disprove his assertion.

The existence of matter is a myth, in the same sense and for the same reasons that, the geocentric model of the solar system and money are myths.

The only reason money has any value is because we agree it does. Money exists in the form of pieces of paper with pictures and writing on it and as coins. But, that form of money is close to being obsolete. For most of us it exists as ones and zeroes located, in the cloud, on a disk drive on some remote server. We get emails or paper statements telling us how much money we have and we all go about our business as if all this was real. What's truly fascinating about money is, even though its reality is more questionable than Berkeley's assertion about the non-existence of the cherry, we conduct our lives and business as if it was more real than a cherry, a book or an atom bomb.

Myth is so deeply embedded in our lives we rarely question it and, sometimes, questioning it gets us into deep trouble.

Why is the matter myth so persistent?

If matter does not exist why is it, when I reach out my hand toward an object, like a glass sitting on a table, I can feel it when I touch it? If it wasn't really "out there" in space time, I shouldn't be able to feel it. Or so it would seem.

Part of the problem with this is the assumptions we make regarding consciousness and how it and the universe are structured. We assume that space–time exists "out there" outside of ourselves and all the objects we see around us. And we assume that all the objects we see, including ourselves and other people are out there too in some kind of huge container or box. This idea makes sense because it squares with our everyday experience. It works well enough to get us through the day but we run into problems when we begin to ask questions about these assumptions. Here's an example: If space–time is really a huge container or box that everything is in, what is the box in? where is it? This is the kind of question a child is likely to ask and, if you asked the question of an experienced and serious physicist or mathematician you'd most likely be told that a) the question is unanswerable or b) the question is childlike and therefore inappropriate. Unfortunately, these answers have nothing to do with the legitimacy of the question.

Here's another question that Avery asks in *Transcendence of the Western Mind*: "Why do space and time shrink and curve and blend into each other?" Special relativity proves that this is so, but the question is why?

Under normal circumstances we don't see this happening. We only see it in situations where things are moving too fast, when mass becomes too great, or when objects become too small. As we approach the speed of light the fabric of space–time begins to tear. As we examine subatomic particles space–time disappears. As we examine unimaginably massive objects, like black holes; space–time turns back in on itself and gravity becomes so dense that light is unable to escape and time comes to a halt.

None of this is speculation. Our mathematics, experiments, observations, and instrumentation has proven it to be true.

We know that, as we accelerate time slows down relative to an observer who isn't accelerating. In this instance the mathematics tells us it's true but so too do experiments. In one such experiment scientists synchronized two atomic clocks. They then put one on an airplane and told the pilot to accelerate as fast as possible for a certain amount of time and then to turn around and come back, once again accelerating as fast as he could. Now certainly he wasn't able to come anywhere near the speed of light but he did accelerate. When they checked the clock that made the trip against the one that stayed on the ground, they found that time slowed down for the clock on the plane in the exact amount predicted by special relativity. They could do this because atomic clocks break seconds down into millions of individual bits. If each second of elapsed time is millions of bits and the total trip time was, say an hour, then the scientists had hundreds of millions of time bits to compare. The extreme accuracy of the clocks made the measurement possible. Time dilation is also regularly demonstrated when subatomic particles are accelerated in particle accelerators. When that happens they approach the actual speed of light.

Here's a question you've almost certainly heard before. If a tree falls in the forest and there's nobody there to hear it, does it make a sound? The reflexive answer is, of course it makes a sound. But, if you think about the question carefully the answer becomes much less obvious. The definition of sound is a particular auditory impression meaning moving air molecules impacting a tympanic membrane or eardrum. If that's the definition then, clearly, the tree in the forest does not make a sound if there's no one there to hear it. Think back to the alarm clock and the bell jar. If an alarm clock goes off in a bell jar that doesn't have any air in it, does it make a sound? Clearly it does not; no air no sound. With the tree in the forest, while the situation is different the logic is the same; no tympanic membrane or listener, no sound. The key difference between the two situations is with the alarm clock and bell jar experiment our observation proves that without any air there can be no sound. With the tree falling in the forest we're dealing with a thought experiment in which we assume no observer exists. The nexus between both situations is the observer.

With special relativity, Einstein illuminated the path which led physicists to quantum mechanics, quantum electro dynamics and quantum chromo dynamics. Quantum mechanics and its sibling disciplines are the most spectacularly successful theories in physics. They have helped us to explain how the world works at the level of atoms and sub atomic particles and they have shown that an observer is integral to understanding what's happening. And, more than that, they've shown that the presence of an observer has a measurable impact on the outcome of physical experiments.

The Particle Dance

Particles like photons and electrons behave differently when they are watched. When Young did his double experiment in the 17th century which proved light was composed of waves he simply shined a light source through two slits. The light waves interfered with each other and formed bright and dark bands. If one of the slits were covered up then only one bright band or splotch appeared on the screen representing the slit.

Today, we're able to control electronic light sources so that it's possible to make the source release just one photon at a time.

If we repeat Young's experiment using such a light source something very interesting happens. Instead of a simple screen we place a detector opposite the plate with two slits cut into it. Then, we turn on the light source so it releases just one photon at a time. As that photon flies toward the plate with the slits it must go through one or the other slit before hitting the detector. The detector can be electronic or it can be a piece of film.

With this set up, if we keep firing photons one at a time, for a long period of time, eventually, on the detector, we will see the same interference pattern we saw when young did his original experiment. Logically this doesn't make sense because we're firing only one photon at a time. As the photon flies toward the plate with two slits logic tells us it must choose one slit or the other so we should see two bright bars or splotches on the detector rather than an interference pattern. But we don't. We see an interference pattern the same as we do when we shine a constant light source of photons through the slits as Young did in his original experiment. What's happening is the photons are interfering with themselves and going through both slits. But, the situation is even stranger than that. When we examine it mathematically we find that the photons are simultaneously going through both slits and through just the right slit and just through the left slit and through neither slit. Mathematically, every possibility is in superposition until, finally we see the interference pattern on the detector behind the slits.

Einstein proved that light can behave as individual particles called photons using the photoelectric effect and yet when released individually in the classic double slit experimental set up they still behave as waves. But here's where things begin to get even more strange. If we set up the experiment again so that our source releases only one photon at a time and we place another detector on the other side of the plate with two slits allowing us to see which slit the photons go through before they hit the far detector behind the plate, the interference pattern disappears and, instead, we see two individual bars of light on the far detector. In the first instance we saw an interference pattern but when we add an observer the interference pattern disappears. The presence of an observer changes the outcome of the experiment!

The first and most obvious question about this is: How do the photons know? Take away the second detector which allows us to know (observe) which slit the photons choose and the interference pattern reappears. Put it back, reintroducing an observer, and the interference pattern disappears, replaced by two distinct splotches of light representing the two slits. Here we have scientific proof that the presence of an observer changes the outcome of a physical experiment. The transcendental question is why? What does an observer have to do with photons moving through slits cut into a metal plate?

Light is strange we already know that. Depending on how we choose to set up our experiments with it, it behaves as if it were composed of particles or waves but never both, at the same time. If all this doesn't make you scratch your head and ask WTF? Then you haven't been paying attention or I haven't done a very good job of explaining what physicists have discovered about light using careful experiments.

So, because it's so weird, let's move away from light and instead examine what scientists have discovered about matter or what we normally regard as stuff.

We know that stuff or matter is composed of atoms and we know that all atoms are composed of the same stuff, protons and neutrons in their nuclei and electrons orbiting around those nuclei. They're like mini solar systems and what distinguishes one atom from another is the number of protons it has in its nucleus. Protons, neutrons and electrons are indistinguishable from each other. If you've seen one proton, neutron or electron, you've seen 'em all. All the matter or stuff in the universe is made of atoms. The only difference between the atoms in steel and strawberry jam is the number of protons in the nuclei of their constituent atoms. Steel, strawberry jam, glass, stars and you and I are made of exactly the same stuff: protons, neutrons and electrons. That's a bit of a bummer but, like it or not, it's true.

Unlike a photon of light, protons, neutrons and electrons have mass. We know how much they weigh. Scientists are able to manipulate atoms and the particles they're made of. Electricity is a flow of electrons.

Since all matter is composed of atoms it should be possible to change one form of matter into another. It is. Scientists do it in the lab with atom smashers or particle accelerators. They can also do it with neutron bombardment. It's even possible to do what the alchemists dreamed of—change lead into gold. The problem is it's so expensive to do it, it isn't worth the effort.

A contemporary of Einstein, Louis de Broglie (1892–1987) like many other scientists was intrigued by the double slit experiment. As he thought about the implications of the experiment an idea occurred to him: If light which everyone thought was composed of waves could behave as if it were composed of particles maybe stuff or particles with mass, like electrons could behave as if they were composed of waves. This idea, based on the work of Einstein and Planck appeared in his doctoral thesis *Research on Quantum Theory*. When his thesis examiners saw it they weren't sure how to respond so they passed his thesis on to Einstein for evaluation. Einstein supported the theory wholeheartedly and de Broglie was awarded his doctorate.

At first blush, de Broglie's idea is crazy. An electron, after all, is a particle with mass which means, among other things, even though it is very small and doesn't weigh much, it can never reach the speed of light. So in that sense it's much the same as a baseball or a chair.

Eventually physicists were compelled to scratch the intellectual itch caused by de Broglie's hypothesis. So they set up the classic double slit experiment but, instead of light, they used electrons. At first they fired a stream of electrons at the slits and what they saw on the detector was two blobs. No surprise there. Electrons, after all, are more like marbles than photons because, like marbles, they have mass. Then they did the experiment again but, instead of a stream of electrons, they fired the electrons one at a time at the plate with two slits and, voilà! They saw the same interference pattern that photons produce when they are fired, one at a time, through the slits. And, just as with the photons, if they used another detector to observe which slit the electrons were going through, instead of an interference pattern they saw two blobs. The question once again becomes, how do the electrons know they're being observed? What does an observer have to do with any of this? What!?

The interference pattern they saw was composed of de Broglie waves and proved that matter or stuff has the same basic structure as light.

Light—again—confounds us—again—but this time muscles its way into an experiment with matter. It seems to be saying, *look at me. I'm everything there is and everything is me.*

If you'd like to get a more visual perspective on all this double slit experiment stuff I recommend you type "double slit experiment" into YouTube or Google. There you'll find video demonstrations and animations that clearly illustrate what I've been trying to explain using just words. The animations starring Dr. Quantum are quite good.

Atoms and particles smaller than atoms exist on the quantum level. Quantum mechanics deals with how particles behave and, what happens, on the quantum level. On the quantum level, certainty is disallowed.

Before quantum mechanics, physicists believed the universe was like a huge clockwork. If you knew the position and momentum of an object's past you could reliably and, with great precision, predict its future position and momentum. This worked extremely well with objects like billiard balls. Newtonian mechanics reigned supreme.

But, when we began to examine objects like photons and electrons suddenly Newtonian mechanics didn't work anymore. Worse than that, we discovered that, on the quantum level, the best we could ever do was talk about probabilities. Certainty about a particle's position or momentum was out of the question.

In his seminal paper on the uncertainty principle Werner Heisenberg said, "The more precisely the position is determined, the less precisely the momentum is known in this instant, and vice versa."

If we attempt to isolate an electron the more we know about its momentum (its mass times its velocity) the less we know about where it is or its position. And, of course the inverse of that is the more we know about its position the less we can know about its momentum.

The reason this is so is, in order to determine where an electron is we have to observe it. To observe it we have to hit it with something like a photon but, when we do that we disturb its position and change it. It's like being in a dark room filled with floating balloons. If we feel around for a certain balloon, the moment we touch it, it moves.

When Einstein learned of the Heisenberg uncertainty principle he said: "*Quantum mechanics is certainly imposing. But an inner voice tells me that it is not yet the real thing. The theory says a lot, but does not really bring us any closer to the secret of the 'old one'. I, at any rate, am convinced that He does not throw dice.*"

Alas, even though Einstein was one of the most brilliant individuals who ever lived he was wrong and Heisenberg was right. On the quantum level He most certainly throws dice, only and always.

We have scientific proof that atoms and subatomic particles exist. Heisenberg's uncertainty principle establishes a real and durable limit on what and how much we're allowed to know about the position and momentum of individual particles. As the physicists watch, the particles dance but, yea verily, how much they're allowed to know about where they are or how fast they're moving is strictly and forever limited regardless of how big and powerful our microscopes and particle accelerators become. On the quantum level, a particle's momentum and position can never be fully known because, even though subatomic particles are stuff, on the quantum level they do not behave like baseballs, tables or chairs.

Reality

Reality exists on multiple levels. The level we're most familiar with is the one we inhabit most of the time, the macroscopic level. That's the level of tables and chairs, baseballs and cars and people. Look around the room you're sitting in. All the objects you see around you are things we associate with macroscopic reality.

Above macroscopic reality we have the cosmos which consists of planets and stars, quasars, galaxies, and black holes among other things. This is the territory explored by cosmologists, astronomers and astrophysicists. What we know about the cosmos is what the people who study it with their satellites and telescopes tell us. Compared to us and the midlevel of reality we inhabit, it's unimaginably big.

Below the midlevel or macroscopic level of reality we have the quantum level. On the quantum level we find atoms, and their constituent particles, protons, neutrons and electrons along with other subatomic particles like mesons, pi-mesons, quarks and photons. We learn about it from physicists.

On the quantum level, careful experiments have shown that reality as it exists on the macroscopic and cosmic levels breaks down or ceases to exist. The double slit experiments discussed above and many of the strange behaviors of light are some examples. So too is the Heisenberg uncertainty principle.

The quantum level underlies the macroscopic and cosmic levels of reality because all the things we encounter on the macroscopic and cosmic levels of reality are composed of the particles we find at the quantum level. Atoms are quantum level objects and all objects, in the universe, are composed of atoms.

The macroscopic level of reality where we exist arises from the quantum level because our world disappears below the quantum level but the quantum level does not disappear in ours.

Here's what I mean. If we want to know the momentum of an object on the macroscopic level, like a car, all we have to do is determine how much the car weighs and multiply that number by how fast it's moving. It's not a big deal. The same would be true for a billiard ball or a snowflake. Momentum equals mass times velocity.

If we want to know the momentum of an electron we're forced into the quantum level. Unlike cars and billiard balls, solitary electrons cannot be seen with the naked eye and they can't be touched with our fingers. We need other particles, like photons, to help us determine where a particular electron is or how fast it's moving. Its velocity or its mass can be determined from its momentum but its momentum cannot be neatly divided into easily distinguishable components of mass and velocity the way it can be with cars and billiard balls. We can know the electron's momentum which, on the quantum level is a fundamental aspect of its existence. But discovering its mass or velocity requires us to carve those parameters out of its momentum and, even when we do that, we can't know which part is mass or which part is velocity.

We understand the world on the level of existence we inhabit, the macroscopic level. When we begin to describe what we see happening on the quantum level in terms of photons impacting electrons and the uncertainty principle, we're using our macroscopic point of view. We do this because this is the only perceptual perspective we have. It's accumulated using information we receive using our senses. But, it's important to keep in mind that our five senses while allowing us to experience what's happening around us also limit the amount of information available. In that sense they operate as valving mechanisms. We can see only a small slice of the electromagnetic spectrum—visible light. Most of it like x-rays, gamma rays, ultraviolet radiation, infrared radiation, cosmic rays and microwaves are invisible to us. It's the same with sound. We can hear only a small part of the spectrum that exists. If we could detect all the information available to us from the universe we would be overwhelmed so valving is as important as having access to what's going on via our five senses.

On the quantum level, the world is very different from the macroscopic world we inhabit. Things we're familiar with or think we're familiar with, like space, time and mass which we can and do use to measure location and momentum—the billiard ball is on the pool table and its moving north at 15 centimeters per second—simply don't work on the quantum level.

The important point to keep in mind about all this is the quantum level underlies the macroscopic level. It's always there even though we can't see it directly. This isn't a guess. It's a scientific certainty.

Science is a very special domain. One reason is that it's objective. Scientists don't mess around with subjectivity or feelings. They conduct experiments and make observations which help them to reach conclusions about how the universe works. Then they publish their work allowing other scientists to examine and replicate what they've done. If other scientists conduct the same experiments and get the same results, then the result of that work eventually becomes recognized scientific fact. The power and value of scientific facts is based on their objectivity. That's the reason the pursuit of scientific knowledge has allowed us to do things like, discover that the earth is not the center of the universe, that everything in the universe is composed of atoms, that atoms are composed of protons, electrons and neutrons and to build bombs with enough destructive power to destroy entire cities with one blast. It's also the reason scientists struggle with what happens when they set up double slit experiments.

Scientists define reality objectively. What they observe and discover has to be observable and repeatable by others, regardless of whether those others are scientists. That is, after all, what objectivity is all about. It's also what makes science so useful, reliable and powerful.

Scientists are interested in physical reality and ascribe to the reigning myth which posits the existence of matter. In other words, they believe matter exists.

In *Transcendence of the Western Mind*, Samuel Avery lays the groundwork for a new myth.

> A new myth will create a world within which relativity theory and quantum mechanics can operate comfortably. But it must also be a world in which we may live comfortably ourselves; if the new myth rejects the concept of matter, as I believe it will, it will have to explain why we seem to experience matter in everyday life. Most likely, what we now call physical reality will be shown to be a special case of a much broader reality.
>
> A new myth will likely say that "reality" is experience in any form, not just dimensional, or "objective," experience. Ideas, thoughts, spirits, dreams, feelings and hallucinations actually experienced will be as real as tables and chairs, only outside of space or time. Images are either in or out of dimensions. *An image in time but not space is a thought; an image in time and space is a perception. An image in neither is just an image.* (italics mine) This redefinition of reality will accompany a shift of assumptions from "consciousness in space–time" to "space–time in consciousness," and does not constitute a separate assumption. It merely puts nondimensional experience on an equal footing with dimensional experience. "Objective" and "subjective" are not separate and not unequal but structurally related elements of the same reality, and composed of the same primal substance.
>
> When reality is limited to the material world, thought, emotion, imagination, spirit, and life itself are reduced to complex neural-mechanical processes, or they are avoided entirely. They are understood as observed from the outside, which is to say that they are understood not at all. Subjective experience makes no appearance in the form of material substance. It is beyond the reach of what can be verified in material terms and beyond the bounds of scientific study. There are scientists who deny that it even exists. Others admit its existence because it is absurd to deny it, but have no idea what to do with it. Most ignore it altogether. It simply does not fit into the world of matter. Even the mention of subjective phenomena in scientific circles produces apologies and embarrassment."[9]

This quote is important; so important you should read it again. It maps Avery's thoughts about how we see the world and how he sees it and it provides hints about what a myth that rejects the existence of matter will look like.

Whether it will, indeed, replace the myth we now cling to even more strongly than we did to the geocentric myth only time will tell. But, my guess is, it will.

CONSCIOUSNESS

Consciousness is a subject about which much is written but very little is known. In some circles it's assumed to be a kind of substance or fluid that conscious beings have located somehow and somewhere in their brains. It might be a flow of electrons or neuro transmitters like acetylcholine. It might also be the result of the firing of neurons as acetylcholine passes between them. It's also possible that it is none of these things.

Years ago, a colleague of mine, Dr. David Zion, and I were discussing management. I was bloviating about one of the latest management trends; about how wonderful and important it was. He took a drag on his cigarette and waved his hand dismissively. Then he asked, "How many books do you think there are on the subject of management?"

I thought for a moment and said, "Most probably thousands."

"People write books about which they know very little or nothing at all." He said. "The more books you can find on a given subject, the less we know about it."

It took me some time to digest what he'd just said and he could see I was struggling with the idea. At the time I was working at Cedars Sinai Medical Center as the director of imaging. Dr. Zion smiled, took another drag on his cigarette and stubbed it out. Then he said, "I want you to go to the medical library, look up the subject of syphilis and tell me how many books you were able to find on the subject."

Cedars Sinai had a substantial and impressive medical library. I went there later that day and asked the librarian to help me find some books on the subject of syphilis. There were none. The reason, of course, was, as Dr. Zion said: "We don't write books about subjects we thoroughly understand, because there's no point to it and, if we did, nobody would read them." It's a lesson I learned well and I refer to it as "Zion's Axiom."

If you Google "consciousness" you'll get millions of hits. You'll also find thousands of books on the subject. We don't know very much about it.

One thing we do know about it is that it's mysterious. People who think deeply about and struggle with defining consciousness try to come to grips with what's known, in such circles, as the hard problem. Susan Blackmore in her *book Conversations on Consciousness* defines it for us: "Briefly stated, the hard problem is the difficulty of understanding how physical processes in the brain can possibly give rise to subjective experiences. After all, objects in the physical world and subjective experiences of them seem to be two radically different kinds of thing: so how can one give rise to the other?"

Another way of framing the hard problem regarding consciousness is to ask; is consciousness something every person has or is it something else; something extra and separate from processes happening in the brain?

One thing that's particularly interesting about consciousness and attempting to understand it is, doing so by definition, requires including subjective experience. Scientists do not enjoy including subjective experience in what they explore. Those who study consciousnesses call the subjective aspects of conscious experience qualia. Redness and sweet smells are qualia associated with a rose. The singular form of qualia is quale. The sound a saw makes when it's used to cut wood is a quale. The problem with qualia of any kind is they're subjective meaning they happen inside the heads of people who experience them. They are not available to anyone else but the person experiencing them.

Color presents some interesting problems. If I show you a red billiard ball and ask you what color it is and you answer "it's red" I naturally assume that you're seeing the same thing I am in terms of color. The problem here is that may or may not be true. It's entirely possible that my subjective experience of red is the same as your subjective experience of blue. We've both learned to call our own particular internal experience "red" but, what's actually happening in our respective heads is available only to us individually. So, if it's true that I see red the way you see blue there's no way for either of us to ever know that. Only the individual who's having a particular sensory experience can know what that experience is. My experience of the color red is mine alone and not available to anyone else. The same is true of your experience of the color red or, for that matter, any kind of perceptual experience. That's why scientists dislike dealing with subjective experiences. This problem has been with us for centuries.

Rene Descartes, (1596–1650) tackled this problem head on. He was a philosopher who understood scientists' aversion to subjective phenomena. After thinking about the situation extensively he declared the physical universe to be a completely separate reality. But he went even further, he said the physical universe was the most important or paramount reality. He didn't deny the existence of a mental or spiritual reality but he insisted that the "substance" of the body and the physical universe be kept separate from the substance of mind or spirit. "By substance, we can understand nothing else than a thing which so exists that it needs no other thing in order to exist."

Well now, there you are you see. Matter exists in its own right, whether or not mind or spirit exists. And, by implication, mind or spirit exists whether or not matter exists.

By establishing this clean separation...forcing matter and mind or spirit into completely separate domains, he avoided any conflicts between the church and science. He established the *primacy* of matter in terms of understanding the physical universe.

After succeeding in doing this he realized that there had to be some kind of connection between mind and matter. So, to explain the soul or what we call consciousness he put it in an organ that lies deep within the brain—the pineal gland. In doing this, whether he wanted to or not, he established the role of mind or spirit as subordinate to the role of matter. He put the spiritual world inside the physical world. Of course we now know he was wrong but at the time, to Descartes, it seemed like a good idea. Following this logic, mental phenomena were somehow supposed to fit in the material world but not the other way around. Physical phenomena existed in their own right with no need of the mental or spiritual world.

It's both interesting and telling that even today, it's pretty much universally accepted that the soul exists within the body.

An additional indication of the extent to which Descartes believed in the primacy of the physical over the mental worlds was the way he described perception.

He said perception has primary and secondary qualities. Primary qualities had to do with properties like shape, extension, mass and hardness. They are objective and absolute and exist, in their own right in the *real* world. Secondary qualities were associated with color, taste and smell. These were not to be trusted because while I may think a slice of watermelon tastes wonderful and is pleasing to the eye, someone else might gag on the taste and perceive it to be repulsive.

The *real* slice of watermelon simply exists in space and time reflecting light and releasing certain chemicals when bitten.

If people disagree about their individual experiences with objects in the material world, the problem is with their perceptions and not with the objects. Material substance is what causes all true experience. Mental substance is, at best, only an approximation and cannot truly compete with the perfection inherent in material substance. In doing this he also separated material substance from religion.

This was and still is a powerful idea; so powerful it has shaped the way we see and interact with the world. It is the foundation of our existing myth regarding the existence of matter. Clearly, Rene Descartes was on to something important.

In attempting to define reality he focused human experience on things that could be manipulated, changed, built; human experience that could be molded and bent to human will—on doing. His ideas became the foundation of technological civilization. That foundation helped us realize it was possible to imagine things that didn't yet exist but could exist if we put our minds to it.

Descartes implied that matter is just out there waiting for us to do with it whatever we might want to do with it. It's not alive and there's no *spirit* associated with its existence. If our attempts to control it fail it isn't matter's fault, it's ours because we lack the knowledge or skill required to do with it what we're trying to do. He also implied that the material world is where the action is; where, if we look carefully, we will find purpose.

That mind set—that vision of the world and how it works established the foundation for what humankind has been able to do since Descartes gave it to us. It was the gateway to the technological revolution. Its foundation allowed us to understand that matter not only exists but that we could manipulate it in ways that eventually wrought everything we see around us; things like billiard balls, telescopes, microscopes, guns, cars, atomic bombs and computers. All things considered—pretty heady stuff.

Under the rubric of Cartesian mythology humankind was to find its purpose in the material world. And, while the mental world exists as a universe parallel to the material world it's certainly not where the action is.

Thinking simply for the sake of thinking was not useful. To be productive we were to concentrate our thinking in ways which lead to the successful manipulation of material objects.

We picked that ball up and ran with it. And, in just under 400 years, we've learned to manipulate matter with such exquisite precision that one scientist wrote the letters IBM using individual atoms.

As impressive as all this is, and it is indeed impressive, our fanatical concentration on the material world has narrowed our *overall* experience of the world. Before we drank the Cartesian Kool-Aid preliterate people everywhere believed in fantastic stories and creatures that were purely imaginary. These stories and creatures were an important part of how they lived their lives. How could they be so childish we ask ourselves now? We think of their beliefs as untruthful and useless and wonder how they could be so ignorant compared to the way we are now.

But, for all we have gained we have also lost much. The narrowing of our collective experience has diminished our ability to appreciate or understand art, music, mystical experiences, tales of miraculous events and mythology itself.

Our preliterate ancestors were able to have experiences outside of perception and doable thought. We are not. We are locked in to the myth Descartes created regarding the existence of matter and the idea that the best and only way to become and remain productive is to continue to focus our attention on what we can do with matter to make our lives better. The things we make have to be bigger, faster, more powerful, and more useful in order to be better than what they ultimately replace.

Today, very few people question Cartesian dualism. The idea that mind and body are distinct and separate seems, for most, intuitively obvious. If someone visits the doctor with a pain in their arm and extensive diagnostic tests fail to reveal a cause, then it's assumed the pain is in their head; they're imagining it and it is, therefore, not real.

Descartes didn't simply pull his ideas out of thin air. His thought processes were careful and rigorous based on introspection and experiments. His ultimate synthesis was clear and logical and it coincided with the ideas others of his generation had about how the world works. His ideas still resonate with modern thinkers in the 21st century.

Descartes' myth was created in the imagination just like any myth is created. His starting point was doubt. To *start fresh* he began by doubting everything—God, religion, himself, the world, matter—everything he had ever believed in. He wanted to dump any mental or conceptual baggage he might be carrying around. His goal was to discover the basis of all truth by going beyond mere belief. He began by reducing experience to consciousness itself.

> Myth is always deeper than belief to those who experience it directly. It is beyond logic and beyond self evidence; it simply is because it is experienced. To those who do not experience the myth directly, it takes on the appearance of belief.[10]

As he continued to pursue this line of thinking he continued to doubt anything that came to mind, even things that were logical. He'd reduced experience to consciousness leaving himself with nothing on which to build a metaphysic. There was nothing to grab onto. Then he noticed the *doubting* itself. As he thought about that he realized *something* must be doing the doubting. That something was, obviously, himself. He realized he must exist in order to doubt that he exists which was the genesis of—"I think, therefore I am." Or, in Latin—"Cogito ergo sum." This was a *fundamental* truth. Then, building upon this truth, he used logic to prove the existence of God, Christianity, matter and all the angels. He concluded that matter is not a given but rather a conclusion based on the existence of self. He starts out with consciousness, finishes with matter and ends up putting consciousness back within matter.

In the battle between Bishop George Berkeley and Rene Descartes to describe how the universe works, Descartes was the winner. His ideas were what humankind needed at the time. They codified our beliefs about how the world works and enabled us to do what we have done. His success was meaningful and spectacular.

But now, the world is different from the way it was when Descartes was alive. "Cogito ergo sum" still obtains but now there are quarks and black holes. We can and do convert matter into energy and energy into matter. We have demonstrated that space and time curve around massive bodies. And, on the quantum level, we have proven that what happens to subatomic particles depends on whether or not someone is watching.

Billiard balls and electrons are forms of matter which we can and do manipulate. They exist on different levels of reality and the ways they behave when we manipulate them offer clues which provide deep insight into the way the universe works.

The existence of matter is still a problem but we now have more tools at our disposal with which to study and probe the question of whether or not it exists.

WHAT'S THE *MATTER* WITH MATTER?

To the average person, taking time and energy to question the existence of matter seems like a fool's mission or, less delicately, just plain nuts.

Matter is everywhere and everything we can see, touch, smell, hear or taste is made of matter. Matter is the *real deal.*

If you look up the definition of matter using Google, you'll get over 36 million hits. Applying Zion's Axiom, and substituting Google for the realm of books, clearly we know very little about it.

I searched Amazon.com and under books, I used the search term "matter". It returned 898,671 hits. Zion's Axiom still obtains even when the search is narrowed to books.

Using Dictionary.com, I found this definition for matter:
—noun

1. the substance or substances of which any physical object consists or is composed: the matter of which the earth is made.
2. physical or corporeal substance in general, whether solid, liquid or gaseous, esp. as distinguished from incorporeal substance, as spirit or mind or from qualities, actions and the like.
3. something that occupies space.

Other dictionaries yielded similar definitions. Matter is the stuff of which everything is made. It is not to be confused with *non-stuff* like spirit, mind, qualities or actions. Clearly, even according to the dictionary, matter can be trusted. As for the other stuff or *non-stuff*, since it isn't matter it doesn't matter. I find that interesting and telling for several reasons.

One reason is the definitions clearly follow what Descartes had to say about matter.

The other reason is that, while looking up the definitions, I was reminded of what R. Buckminster Fuller said in *Synergetics* about dictionaries. He made the observation that when we go to the dictionary (this was in the early 70's when going to the dictionary meant picking up a physical book) to look up a word, we often find ourselves lost in the dictionary long after we found the definition we were looking for. The reason that happens, he said, is that the dictionary is essentially a catalog of the universe. I think he was right. I often found myself lost in the dictionary long after I'd found the definition I was looking for. Now, not so much; the internet and Google, while certainly providing instant gratification when looking up the definition of words, have stripped away the delicious pleasure of meandering through the pages of a physical dictionary and getting lost in what was there to be found.

It's also interesting and telling that when you begin to focus on the word matter in an attempt to truly understand what it means and, ultimately, apprehend what matter is you find yourself in the same kind of intellectual/epistemological quagmire associated with doing the same thing for the words time, space and now.

We're all pretty sure we *know* what the words time, space, now and matter mean until someone asks us to explain what they mean; especially if that someone is a child. I find that very interesting.

Let's do what we did before. Let's see what the physicists know and have to say about matter.

Matter occupies space. The space occupied by matter is unavailable to any other matter until or unless the matter originally occupying it is moved out of the way. Put differently, no two objects can occupy the same space at the same time. We see the truth of this demonstrated every time we see a stationary billiard ball get hit by a rapidly moving billiard ball on a pool table. If you've ever been in a car accident, you know, in a very immediate and personal way, that no two objects can occupy the same space at the same time.

Matter comes in various forms. We normally think of it as hard. Things like billiard balls, bowling balls, bricks, stones and mountains come quickly to mind. Then, there are things like bread dough. Dough is matter too. It's soft and can be shaped but, it's still matter. Smoke is matter too. If you see smoke rising from a cigar sitting in an ashtray you can move your finger through it. What really happens when you do that is you push the smoke particles out of the way. Your finger doesn't really go *through* the smoke. It's the same deal with an airplane pushing through a cloud. Smoke and clouds are composed of millions of tiny particles held together by electrostatic forces. And, even if we don't think of them as matter in the same sense we think of stones and mountains, they are matter nonetheless.

It's not possible for physicists to detect matter directly for the same reasons we can't detect it directly; they're people just like us. What physicists can do, however, is detect the presence of matter through mass. An object's mass is a determination of how much matter that object has. The more matter it has the heavier it is. On the atomic and subatomic levels what makes one atom heavier than another is the number of protons and neutrons in its nucleus. An atom of the gas hydrogen has one proton in its nucleus with an atomic weight of 1 while an atom of lead has protons and neutrons in its nucleus which add up to an atomic weight of 207. All matter is made up of atoms and most of the mass of any object is the result of the number of protons and neutrons in its constituent atoms.

Physicists measure an object's mass by determining the extent to which it resists acceleration. If you want to accelerate an object you have to push it or pull it. If you push a heavy object, like say a box filled with lead shot it will not move as fast as when you push a similar empty box. In physics—speak the heavy box will not accelerate as fast as the empty box and it will not change its velocity as rapidly as the empty box. Of course, we all know this because we've had the experience of trying to push something that's heavy versus something that's light. We don't think much about it because that's just the way things are.

What's really interesting about this, however, is we don't know why this is so. I know what you're thinking; Well, DUH! The box with the lead shot is heavier. And, while that's certainly true, it doesn't explain how or why the heavier object is able to resist acceleration. You may also be thinking about the friction caused by gravity that must also be overcome in order to accelerate either box but, here's the deal with that. If we were on the space shuttle, in orbit, and we had the same two boxes floating in front of us, while they'd both be weightless, it would still take a lot more force to accelerate the one filled with lead shot than it would to accelerate the empty one. Try to imagine it. They're both weightless, yet it's much harder to get the one filled with lead shot to move, and it would also be much harder to stop it or slow it down once we got it moving. If the one filled with lead shot were to collide with someone, it would hit him much harder and move him much farther than the empty box would. The question about all this is why? The answer is, nobody knows. It's almost as if heavier objects are somehow able to resist acceleration by extending invisible claws that cling to empty space. I don't know about you, but I've never seen no stinking claws. It's another mystery like wave particle duality.

Massive objects resist acceleration but not simple velocity. Go back to the space shuttle in your mind. If you push the heavy box with enough force you'll overcome its resistance to acceleration and, once it's moving, it will continue to move at whatever velocity it has attained. It will not slow down until or unless it collides with something like another object or a wall and if you were on a spacewalk with the box if it attained a velocity of, say, twenty feet per second after you pushed it, it would continue moving at twenty feet per second forever until or unless it collided with another object. Nobody knows why.

Before the twentieth century most physicists believed space was absolute. In other words, they believed every point in space was fixed and never changed. Thinking about space as absolute made it easier to imagine massive objects extending their claws into it to resist acceleration even though nobody ever saw any claws.

Today the problem is more difficult because most physicists believe space is relative rather than absolute. They think of space as having meaning only as a relation between objects. Here's what I mean. You can say the lamp is two feet from the wall and six feet from the sofa but it's nowhere in relation to space itself. Objects do not occupy or pass through fixed points in space but, if that's true, what is it that they cling to when they resist acceleration? Why do they slow down after we push them and what does that have to do with how much matter they contain?

We use mass to detect matter but the only way we can measure mass is through changes in its motion. Part of the problem with all this is definition 2 draws a bright and unambiguous line between matter and *actions*.

There's one other way physicists can detect matter—gravity. Newton believed that massive objects are surrounded by a gravitational field that attracts other massive objects. That's how he explained why the moon and the earth influence each other's motion and they do it instantaneously even though there's no physical connection between them. The problem with that explanation was it required *action at a distance*. But, *action at a distance* was and still is forbidden by classical physics and Newtonian mechanics.

Einstein cleaned up the action at a distance problem by abandoning the idea that gravity was a force. He said that massive objects do not *attract* each other. What they do instead is curve space-time in their vicinity. As objects move through this curved space-time they only appear to be moving in curves. In other words, it only *looks like* they're being attracted by a force. Regardless of which way you look at it, gravitational fields are associated with massive objects and indicate the presence of matter. The problem with all this is we measure gravity in terms of acceleration too which means the only way for us to detect matter is in terms of actions. Definition 2 and Descartes insist that matter and actions are two very different phenomena. They're not connected in any meaningful way. Matter just is. Actions have to be *done*.

There's another thing about mass that's interesting if you think about it carefully. Just looking at an object doesn't tell you anything meaningful about its mass unless, of course, you've had previous experience with the object. Here's what I mean. If you were to see two identical boxes sitting on a table and one was filled with lead shot while the other was empty; just looking at the boxes would tell you nothing useful about their mass. The only way you'd be able to discover the difference in mass between the two boxes would be to move them.

If we put aside what dictionaries and physicists have to say about matter we know that even though it's not possible to measure mass without resorting to some kind of action we can *feel* it. Objects that have more mass are heavier than objects with less mass. They're harder to lift, push or pull. If you've ever moved to a new house or apartment and gotten some friends to help you know what I mean when I say you have an intuitive sense that objects have mass. Carrying a refrigerator up a flight of stairs drives the point home with no ambiguity. It's mass can be felt directly and you learn quickly that such objects have inertial resistance and they tend to accelerate in a gravitational field. If you let go of the refrigerator it will fall to the bottom of the staircase. But, it's important to keep in mind here that what you're *feeling* as you move the refrigerator is its mass not its matter. Mass and matter are not the same.

The source of our intuitive sense of mass is tactile perception. The trouble with that, from a scientific perspective, the way a physicist would look at it, is tactile perception cannot be measured in any precise or meaningful way. No one would argue that tactile perception exists. It exists in the same sense that visual, auditory, olfactory and chemical perceptions exist. But, there's no way to objectively quantify it. It falls into the category Descartes described as a *secondary* quality and belongs alongside things like smell, taste, sound, and color; interesting but essentially untrustworthy and unimportant.

Yet another strange property of mass is, as Einstein showed, it increases as an object accelerates. Any object, with even the slightest amount of mass, when accelerated gets heavier. If that object continues to accelerate to the point where its velocity approaches the speed of light its mass becomes infinite which prevents it from attaining light speed. Objects with mass, even if they're subatomic particles, can never attain light speed.

The units physicists use to measure mass are meters and seconds but the measurement is not as straightforward as when they measure velocity or speed. Velocity is relatively easy to measure and understand. If it takes an airplane two hours to travel 1000 miles its speed is 500 miles per hour. Velocity equals distance divided by time.

Mass, however, is detected through acceleration. So, unlike simple velocity, it must be measured in terms of meters per second per second or, put differently, the change in velocity over a given period of time. It takes a Porsche Turbo 3.5 seconds to get from 0 to 60 miles per hour. If we poured 500 pounds of lead shot into the space behind the seats it would take the Porsche longer to get from 0 to 60 miles per hour, maybe 4 seconds. This is important because it is this second time component that *defines* mass. Take that second time component away and there is no mass because without it there's no way to detect it. We know the Porsche's mass increased because it takes it longer to go from 0 to 60 miles per hour.

If we think about all this just a little more carefully we come to understand that without that second time component there wouldn't be any tactile sensation either. Here's the reason. If you were sitting in that Porsche traveling down the road at a steady 60 miles per hour you wouldn't be able to feel the velocity. Even if you were traveling at a steady 100 miles per hour it wouldn't feel any different from 60 miles per hour. We can't feel steady velocity. Think about the last time you were on an airplane. You were probably traveling well over 300 miles per hour but you didn't feel it. The only thing you can feel is when the velocity changes. If you were in the passenger seat of the Porsche Turbo when the driver accelerated from 0 to 60 in 3.5 seconds you'd feel your neck snap, your eyeballs press hard against their sockets and your head slam into the headrest. And if, after reaching 60, he slammed on the brakes, you'd feel that too, only your chest would be slamming into the seatbelt, your head would snap forward rather than backward and your eyeballs would be trying to exit their sockets. We can't feel simple velocity; only changes in velocity. That's what the second time component—*the second—per second* is all about.

The reason any of this is important is; that second time component is more than just a second time component. What it *really* is—is another dimension.

When we measure anything we identify as material, we do so only in terms of mass. We're able to feel mass and when we do measure it we do so as a second time component. Looked at from this perspective it isn't at all necessary that mass be a component of matter. And, even though that's not the way we normally think about mass and matter doing so doesn't violate any laws of physics or science. The physics and science work perfectly well even when we separate mass from matter.

One other mystery associated with mass that still has physicists scratching their heads is—no one understands why objects and subatomic particles have it. One of the things scientists hope to discover with the Large Hadron Collider or LHC operating at CERN is the Higgs boson or God particle. The theory states that if the Higgs boson exists it may be what's responsible for the mass we're able to detect in macroscopic objects and subatomic particles. We'll just have to wait and see.

Careful review of what we do know about mass doesn't give us a reason to doubt the existence of matter. If what we know about physics pertained only to the macroscopic or mid level of reality—the level of baseballs, hotdogs, apple pie and Chevrolets, then we might as well hold fast to the idea that matter really does exist. The idea gives meaning to what we do and what we experience on the everyday level of existence. The problem we're faced with, however, is when things become extremely large, like galaxies and the cosmos, extremely massive like black holes, extremely small like subatomic particles or when they begin to approach the speed of light, that same physics either breaks down or doesn't work at all. One level above or below the level of reality on which we normally exist the physics we've come to know and rely on either doesn't work or becomes so iffy it can't be trusted.

If our goal is to truly understand physical reality, like it or not, we're required to include the realms of the extremely large, the extremely massive, the extremely small and the extremely fast because even though we rarely encounter them, they do indeed exist.

Let the weirdness begin

Before the twentieth century physicists, as I've mentioned earlier, were a relatively smug and self satisfied group. They believed they'd pretty much figured out how the physical universe worked and they were worried that, in the future, there wouldn't really be much left for them to discover or do. Looked at in terms of what we now know such unmitigated hubris is laughable. At the time, they truly didn't know any better. It wasn't until the beginning of the twentieth century that scientists began to explore what happens at dimensional extremes. By the turn of the century they'd begun to develop instruments, techniques, and the curiosity required to explore phenomena like distant galaxies and atoms.

In the seventeenth century, scientists believed the physical universe could be completely understood by examining what it looked like on what we now know to be the macroscopic level of reality—the world of tables and chairs. In the early 1800s while they almost certainly had apple pie and later baseballs and hotdogs they didn't have any Chevrolets. They believed that if you looked out into space you'd see bigger tables and chairs and if you looked through a microscope you'd see smaller tables and chairs. Macroscopic reality extended to the cosmic and the atomic levels.

The universe was composed of perfectly square dimensions. Time had no connection to space and flowed smoothly, at the same rate, everywhere. The discipline of physics was concerned with how matter behaved and moved within this universe. Consciousness had nothing to do with how the universe worked and was unnecessary in understanding how it worked.

This was an extremely narrow view. The problem, of course, was they had no idea of how big the universe really is. As far as they knew, there were no galaxies beyond the Milky Way and atoms were, at best, hypothetical.

Before the twentieth century, physicists believed the secrets of the universe, including life itself, would one day be revealed as completely predictable physical processes. Smug and hubris almost always go well together.

They also believed that energy flowed smoothly; that it existed in a perfectly continuous and divisible state.

In 1900, Max Planck was studying a phenomenon called black body radiation. One reason he was studying it was to determine the best material from which to make light bulb filaments. To his surprise and dismay, he discovered that energy *does not* exist in a perfectly continuous state. Planck was extremely conservative so he found his discovery to be enormously disturbing. He'd fully expected to discover that the reason materials like iron glowed either red or white when heated was that energy does exist in a perfectly continuous state and that physical principles already in existence could readily explain the phenomenon. Instead he found that energy exists in the form of tiny bundles he called *quanta*. For Planck, this was quite upsetting. He wasn't at all interested in discovering new phenomena regarding how the universe worked. Initially, he thought his observations were wrong. And, more disconcerting than anything else, his observations didn't agree with well established theories. As he struggled with the information his theories were yielding, against his better judgment he applied Boltzmann's statistical mechanics of the second law of thermodynamics *entropy* to his black body radiation law, even though he had a strong aversion to doing that. His recourse to using that approach was, for him, an act of despair. Of it he said, "I was ready to sacrifice any of my previous convictions about physics."[11]

Initially, he believed what his observations revealed about quantization, were nothing more than purely formal assumptions. That conjecture is incompatible with what we now know about particle physics.

Planck's discovery of the quantum is considered by most theoretical physicists as the birth of the discipline of quantum mechanics. Its implications reach far beyond what he was trying to do at the time and have changed the way we see the world along with our understanding of how the universe works. In order to embrace the implications of Planck's discovery it is necessary for whomever whishes to do so, to be willing to *walk through the looking glass* just as Alice did in Lewis Carroll's famous book.

Prior to the development of quantum mechanics, physicists believed that subatomic particles, like electrons, behaved pretty much the same way as billiard balls. That idea was comfortable and made sense. But, quantum mechanics revealed that it wasn't so. The energy of a billiard ball moving across a pool table is smooth. An electron does not move anything like a billiard ball. Quantum mechanics revealed that, instead, it *leaps* from one point to the next without ever occupying the space between leaps. On the quantum level space, as we understand it on the everyday macroscopic level, doesn't exist. Once you walk through the looking glass, weirdness abounds. You're forced to abandon ideas that lend comfort to understanding what's going on; ideas like space, classical Newtonian mechanics, and certainty. Probability, not certainty, reigns supreme on the quantum level. Space, time, and mass become undistinguishable and the more you know about one, the less you're allowed to know about the others. What made this so disturbing was it became apparent that there is a physical limitation on our ability to locate particles in space and time and, to make matters worse, that limitation is built into the universe itself. Put differently, there's no way around it. Regardless of how big, powerful or precise your instruments are or become, you are forever forbidden from knowing exactly where or when a particle exists. At the time, this was weirdness on a gargantuan scale. And, worse yet, almost as if to rub salt in the wounds already opened up by quantum mechanics, it became apparent that the very act of observing these particles affects their behavior! It turns out that *we*, as conscious human beings, are intimately connected with physical reality. Physics had always been the hardest of the hard sciences. But, on the other side of the looking glass, weirdness becomes and remains part of the package.

When Einstein, in 1905, formulated his special theory of relativity, in addition to the many odd things he discovered as discussed earlier, he also discovered that there's something about the way the fabric of space–time is structured which imposes a very real and durable limit on the speed at which objects can travel. That limit is the speed of light. Additionally, he discovered that light always travels at the same speed relative to everything. The speed of light *establishes* the universal speed limit.

These discoveries by Planck and Einstein were and still are weird. But, in addition to that and of greater importance, they showed that the idea we held so near and dear, of the universe as a *box* filled with lifeless matter and some observers couldn't possibly be right.

What's interesting about all this is, more than 100 years after what Planck and Einstein discovered, we still cling to the idea of the universe as a *box*. We do that because the idea is comfortable and seems like home. The problem we're faced with is we *have* stepped through the looking glass and, because we have, even if we're not but especially if we are a physicist, we're forced to examine much more carefully just what the box is and, whether we like it or not, what we are.

The pursuit of science expands what we know about how the universe works and further defines our relationship to and with it. Once that knowledge has been acquired, even if we want to, we cannot ignore it.

Meditation

Meditation does not fit well within traditional Western culture. This is especially true of American culture. For most Westerners the idea of meditation is, at best, iffy. Questions arise like, how do you do it? What's supposed to happen when you do it? What's the point of doing it? Does it work? How will I know if it's working?

These are all legitimate questions. As much as anything else, they reflect the traditional Western/American mindset. They also indicate how different our culture is compared to Eastern cultures where asking such questions would seem unnecessary or out of place.

Culture and myth reinforce each other. The Western mindset has a Cartesian bent often associated with phrases like, *when I see it I'll believe it*, or *if I can't see it, touch it, taste it, smell it or hear it, it isn't real.*

We like to *see* things for ourselves. If we watch someone meditate, we see him sit down and close his eyes for 20 to 30 minutes after which he opens his eyes and walks away or begins doing something else. It's hard to extract useful information about what happened because it all happened inside the head of the person who was meditating. Even if he describes what his experience was like, we're faced with the same dilemma we had earlier when we examined consciousness and qualia. My red might be your blue.

I know a little bit about meditation because I've been doing it for 35 years. The technique I use is Transcendental meditation. But there are others. Meditation falls broadly into two categories, contemplative and concentrative. In the contemplative approach, you focus your thoughts on one thing, like a koan. In Zen meditation koans are paradoxical questions or statements. Here's an example: *What is the sound of one hand clapping?* The idea is to force yourself to abandon normal logic by trying to answer an unanswerable question and thereby gain enlightenment.

In the concentrative approach you focus your attention on one object, like a candle flame, in an attempt to empty your mind, which then leads to enlightenment.

Other techniques require sitting in certain positions, like the lotus position, which, for most Westerners, is extremely uncomfortable. Yet others involve movement like the dancing that whirling Dervishes do or the controlled slow movements performed by practitioners of yoga and Tai-Chi-Chuan.

Transcendental meditation uses a mantra which is a word that has no meaning. It's more a sound than a word and, after closing your eyes, you *experience* that sound by listening to it in your mind. Transcendental meditation TM, was created by the Maharishi Mahesh Yogi and brought to America in the 60s specifically for our Western culture. It's done sitting in a comfortable chair. It's neither contemplative nor concentrative.

Westerners believe the purpose of meditation is looking inward or introspection. And while it can be used for introspection the Eastern approach is the opposite. It's a method used to detach from self or the "I" that's always present and thus becomes a different way of being. The central problem for anyone attempting to do that is the self is very strong. The self arises in order to do. It always wants to be present. Attempting to detach from it causes problems that interfere with meditation. Meditation is *not* doing. The self wants to participate, it wants to *do*.

What you're really trying to accomplish when you meditate is to *be* different; to see the world the way it really is by detaching from it and yourself. It's easy and it's very difficult at the same time. You close your eyes and try to shut your thoughts off. Then you realize you're thinking about trying to shut your thoughts off and then, thinking about thinking about trying to shut your thoughts off; not so easy after all. It's a kind of dance that's very difficult to learn on your own; not impossible but very difficult. If you decide to try meditation I recommend finding someone who can teach you how to do it and who can help you find answers to the many questions that will arise as you attempt to learn to meditate.

I believe meditation is important for many reasons, not the least of which is, it can be used to help understand aspects of existence and how the universe works in ways unavailable using any other approach. It has also been proven to have significant and long lasting health benefits.

One of the problems for anyone attempting to learn meditation is labels. From the moment we're born, we begin the process of learning about and then attaching labels to everything we encounter. For most of us the earliest labels we acquire are mommy and daddy. We learn who mommy and daddy are and why they're important and we learn to depend on them. Initially, they help us acquire more labels like house, bed, car, school, food, doctor, aunt, uncle, grandmother, grandfather and on and on.

As we get older we learn to acquire labels from other sources. The labels help us to make sense of the world. Once acquired the labels become part of the database we carry around that helps us understand how the world works and what it is. The labels are important because, without them, we would have only chaos. Absent the ongoing process of acquiring labels, everything we encounter would be new and unfamiliar every time we encountered it. We'd have to figure out what a shoe is every time we saw one. In this sense the labels become a shortcut to both understanding and dealing with what we encounter.

Even though acquiring labels is mostly automatic, it is an arduous process. Label acquisition exerts a significant and ongoing effect on the way we perceive time. When I think back to when I was very young, I remember that time seemed to pass at a much slower rate than it does now. Summer vacations from grammar school were wonderful and seemed to last forever. The amount of time that existed between signal events like summer vacation and Christmas seemed interminable. And the dot on my personal time horizon that represented adulthood appeared to lie eons into the future. The reason is when we are very young, by definition, everything is new. Everything we encounter is something we've never encountered before. So, we're faced with the task of first determining what it is we've encountered and then acquiring and attaching a label to it. This process is both arduous and time consuming which explains why time seems to move so slowly when we are very young. There's so much to learn that the task, at times, seems impossible. For some of us, as we get older, we come to realize that the more we learn the less we really know.

As we continue to move along our own personal time line, we acquire more labels. As our store of labels increases, our need to attend to the task of figuring out what's going on in the world requires less time and effort. We no longer have to expend time and effort trying to figure out what a shoe is. We know what it is and we have many labels in the subcategory of shoe that include boot, sandal, flip-flop, slipper and high heel.

Eventually, we acquire so many labels that our perception of time reverses. For most of us this happens slowly. We may begin to notice that summer vacation seems to go a little faster than it used to. We get surprised that, before we know it, Christmas will be here again. Days aren't as long as they used to be and weeks, months and years begin to slip by at a pace that, for many, becomes disconcerting. As we get older, the content of our consciousness increases at a decreasing rate. We have tons of labels.

For most of us, most of the time, labels are good. They provide a framework for knowledge. They can and do, however, get in the way when we try to meditate and when we try to bring expanded understanding to certain aspects of how the world works.

Done properly, meditation allows you to transcend the material world.

One important thing you're trying to do when you meditate is remove the labels. It's easier said than done because you have to really think about that, and what it means, before you can attempt to do it.

Here's what I mean. If I'm sitting quietly reading and I suddenly hear a loud chittering sound outside what I think I'm hearing is a small furry creature with a bushy tail, a cute face and black eyes that look like big oil drops—a squirrel. The moment I hear the sound, since I've heard similar sounds before, I attach a label—*squirrel*. For me, as for most people, the labeling process is so ingrained and automatic that I'm not mindful of it. I just do it. But, here's the problem with that. The chittering sound is new. I've never experienced it before and never will again. What's not new is the label—*squirrel*. I hear the chittering, attach the label squirrel to it, and file it away with other similar experiences. This helps me to make sense of what's happening but it also makes me stop paying attention to it.

The chittering is *not* a squirrel, it's a sound. When I heard it, I *envisioned* a squirrel which caused me to attach the label. I know that if I were to get up and go to the window I would most probably see a squirrel. But, that doesn't make the sound a squirrel; it just confirms the label. To fully experience and appreciate the chittering sound, I have to remove the label and detach myself from the image of a squirrel. I have to look at what is. And that's the *hard* part.

Everything you experience is new and different every time you experience it. I know that sounds crazy but you have to think about what it means. You get in your car and drive to work. On the way you may think to yourself, same old thing, driving to work. That's the *label* and the *label is* the same old thing; the experience, however, is not. You will never have that exact same experience again for the same reason you cannot put your finger, in the exact same place, in a moving stream. The stream is always moving; so is time.

We can and do attach labels to every sensory and non-sensory experience we have because that's how we get through the day. If you make a conscious effort to remove the labels associated with any familiar experience you'll be able to recognize its newness. One way to think about what this means and how important it is, is to think back to the last time you saw someone do a really good magic trick. The immediate reaction is Whoa! How did he do that? I remember the first time I saw David Copperfield fly. I was stunned. I had no labels to help me understand what I was seeing. And, it was doubly baffling because I've been intensely interested in magic for most of my life so, even with an almost encyclopedic knowledge of how magic is performed, it was difficult for me to make any sense of what I was seeing. That's the nature of the experience you can expect *if* you succeed in removing the labels from any familiar occurrence or event. That feeling of not being able to explain what you're experiencing is a glimpse into the chaos; a glimpse of what it would be like if you had no labels.

The self is interested in doing and in succeeding by doing. This is true for any endeavor including meditation. What makes the effort paradoxical is, in order to truly succeed, you must detach from self and from the world. Stripping away the labels is a good place to start because, to the extent you succeed, you will be able to look at what is as it is. You'll be able to pay attention to what is without the labels getting in the way but it is not easy to do.

I recall reading an account of a student of Zen meditation asking his roshi (teacher) what he should do when thoughts constantly float up from his consciousness and interfere with his attempts to detach from self and the world. His roshi said, "You can let the thoughts come but you do not have to serve them tea."

You can try to learn meditation on your own. Sit comfortably in a chair. Close your eyes and try to clear your mind of thoughts; to strip away as many labels as possible; to look at what is. You will be able to watch your mind trying to do that and you will, most probably, be surprised and annoyed at how difficult it is to do what you're trying to do. You will be serving tea to the thoughts that arrive and demand attention. That is the self, grabbing. Try to watch that and try to watch yourself watching that. It isn't easy to do but it is worth the effort. With practice and with a competent teacher your chances for success become better. Eventually you will learn that techniques exist which can cause the self to become tired of grabbing or to become distracted from grabbing allowing you to glimpse what is. When that happens, you will eventually be able to glimpse the chaos.

The reason meditation is important is, done properly, it opens a pathway for understanding what cannot be understood while holding on to the labels and remaining attached to the world and yourself. It allows you to truly understand that the chittering sound I mentioned earlier is not a squirrel or that the droning sound you hear outside, in the sky, is not an airplane.

Learning to see *what is* is difficult and painful because it requires just as much effort as you expended in acquiring the labels you have and becoming aware of and attached to yourself and the world. To truly understand how the world works you must be willing to detach yourself from the world—to truly see what is. You must be willing to look into the chaos as it churns and roils and swirls spewing thoughts, ideas, pieces of the world and objects. It's always there just underneath what appears to be the objective world we live in; the middle latitudes of reality. The problem is, if you see it, you are not you because, in order to be able to see it you must detach from your self and from all the labels that help keep everything calm and familiar. If you see it you will see only shards and pieces—glimpses and because you were not *you* when you saw it you will not remember what you saw—when you were detached.

All this is very counterintuitive. It purposely runs against the grain of the Western/Cartesian mindset. What's important to keep in mind about this is, that's the point. At bottom, meditation is a different way of being. It's a way of bringing expanded understanding to the universe and how it works by establishing a different relationship to it through the attempt to detach from the universe and from yourself.

One way to bring some understanding to why detaching is so difficult is to think about my favorite definition of the word universe from R. Buckminster Fuller's *Synergetics. For each individual the definition of universe must be everything that isn't me plus me.* It's difficult to detach from everything that isn't you plus you; difficult but not impossible.

The next section starts with a passage from Samuel Avery's *Transcendence of the Western Mind*, in which he explains meditation and some of the problems inherent in trying to do it.

Look past self

> Now look again at what is, Follow the breathing, or recite your mantra, or do whatever you do to calm the mind. As always, do not try to make anything happen. This time, try to hear what you are hearing and feel what you are feeling without thinking "I" hear this, or "I" feel that. Just listen to the sound itself. Feel the feeling without the feeler. You cannot stop the "I" from arising, but you can watch it arise. It does not have to be there. It wants to be there, but it does not have to be. It will be hard to see past it because it is very strong, as strong as you are, and it will want to be there at least as much as you want it not there. Even the wanting of it not to be there is what it is.
>
> *The self arises from being in order to create doing* [emphasis mine]. When you feel a pain in your lower leg, the self wants to move it somewhere else, and thoughts arise as to where and how to move it and why you should do it right now instead of waiting until you said you would stop meditating. The self is not satisfied just looking at the pain, and this is why it is so difficult to be without doing, and why when you do it you cannot do it at all.
>
> But you can do it. Do not try to make it happen; let it happen. Do not even let it happen; just watch what is happening. There is a place that is breathing in and breathing out; breathing in and breathing out, and you may watch it anytime. It is not you breathing in and breathing out, but it is the only one that you will see. The mind wants to grab it and say "it is only me breathing," but forget about that. Let it grab and release. The mind will tire of grabbing, but the breathing will continue.[12]

One reason it is so difficult to see chaos is chaos is not in time. That's also the reason you don't remember having seen it. When you see it, you are not you. Your *self* shields you from the chaos. Detaching from your *self* allows you to see it but, once detached, you are no longer your *self*. This seems contradictory because meditation done properly allows you to see something that doesn't exist in time. Thinking about meditation is not the same as meditating. Meditation is what gets you there.

Samuel Avery says: "The chaos is consciousness. The labels are the world." These two statements fly in the face of the way most of us understand the world and how it works. They also lead us directly to the precipice of the abyss of Solipsism.

Most people are not familiar with the word Solipsism. A Solipsist believes that only he exists and that everything he sees of and in the world happens in his mind. His perceptions create his reality and beyond his perceptions nothing exists.

The initial reaction of most people when they learn what Solipsism means is to reject the idea as being ridiculous. One reason is Solipsism implies that, if everything is in me then, in addition to all the *things* in the world being in me, other people are in me too. They don't have consciousness.

If it's true that chaos is consciousness and the labels are the world then it follows that the world arises out of consciousness. But if Solipsism is true then the question becomes who gets to have consciousness? Is it me, or you or someone else? Westerners believe everyone *has* his own consciousness. This makes the very idea of Solipsism ugly and unacceptable and it is one reason it's referred to as an abyss.

Even though all this is true solipsism is problematic because, when it's examined in a highly disciplined way, applying the full force of logic, it becomes impossible to deny. Cold, hard scientific examination points unambiguously to solipsism. Scientists, perhaps more than non-scientists, don't like to think about solipsism. The intellectual disciplinary domains most often associated with the exploration of solipsism and its implications are philosophy and metaphysics.

The idea of solipsism is unsettling because the most reliable and powerful means we have for exploring the universe is science and the scientific method. When we apply logic and the scientific method to explore the implications of solipsism the result we get is the implications are true. Let us leave solipsism for now and return later after we've explored some detailed explanations for how and why we experience the world the way we do.

One thing meditation reveals is there are many different kinds of labels. The most elemental labels are associated with experiences we have like seeing, hearing, smelling, tasting, touching, and thinking. We understand what these categories are and how they're structured. We also understand that they are only a subset of all the labels available. They don't include things like fear, love, hate, envy or desire but, at the level of the sensory domains or perceptual realms most would agree that seeing, hearing, tasting, smelling, touching and thinking are highly structured. The *structure* of the first five of these implies for us a world that is much larger than what any one of them contains. If I hear something I get the immediate impression that more than just that sound exists. I assume that *something* is making that sound and that there's a world in which that *something*, whatever it might be, is in.

When I heard the chittering sound I described earlier, I did not hear a cute furry little animal with a bushy tail and black eyes that look like oil drops. I did not hear my back yard or the grass and trees and swimming pool in my back yard. I envisioned or thought these things but I did not hear them. I envisioned them because I have seen squirrels making chittering sounds in my yard. I have offered them peanuts and had them come up to me and take the peanuts from my hand. The direction the sound was coming from told me that what I have seen and touched before was in my yard. The sound made me think of things like cute, furry, bushy tailed, golden brown but I did not actually experience a squirrel in my yard. I assumed he was there making the chittering sound and that was the reason I was hearing it. At that moment, for me, there was an object and a world. But the important part of this little story, the part I want you to focus on, is all I *really* knew was there was a chittering sound.

Given our Western sensibilities and Cartesian mindset, we feel almost forced to assume there is a squirrel and a world in which that squirrel exists. Absent that assumption it would be difficult to make any sense of what was going on. I'd have no way to explain why, on the previous day, I saw a squirrel making that same sound in my yard.

If we don't think about this situation too carefully we assume there is a squirrel in the world making the chittering sounds and evoking the visual impressions of cute, bushy tailed and golden brown when we happen to be looking or touching. We also must assume that he's there whether we're looking or not. We go even further than that. We assume the squirrel exists whether or not we have any kind of perceptual experience of him; that is even if we don't see, hear, feel, smell or taste him we assume that he's still there. The stuff he's made of is more fundamental than consciousness and, even though the only way we can perceive him is through consciousness we have to say that he is there all the same. That *belief* that he is there is what holds the world together for us. And it appears to be very reliable because no matter how we choose to test it, it works.

Let's say I hear the chittering sound but I'm not convinced that there is a squirrel making it. So I get up from my chair and go to the window and there he is. He stands up on his hind legs while gnawing on a date from one of the palm trees, all the while making that loud chittering sound. I'm experiencing a visual and aural perception of a cute furry, golden brown bushy tailed creature. And, while that's true, there still isn't any strict proof of a squirrel in the world. The chittering sound is coordinated in space with the visual experience I'm having and, even though there is no proof of the squirrel's existence my assumption is holding up well. Let's say I'm still not convinced so I get some peanuts and go outside. At first the squirrel runs away when he hears the door open but, since he knows I might have peanuts he doesn't leave the yard. Instead he stands up on his hind legs and watches me. I squat down and hold out my hand in his direction. He hesitates for a few seconds and then scurries over to me, reaches into my hand and snatches a few peanuts after which he runs a few feet away and watches me while he eats them. When he snatched the peanuts I felt his little claws and the fur on his front legs. There's still no strict proof of a squirrel outside of consciousness but the tactile, visual, and auditory realms or domains of consciousness appear to be dimensionally coordinated. The strong implication of all this is, there is a world outside of consciousness that causes the dimensional coordination of what I saw, heard and felt. This is what most of us assume most of the time. The world and the dimensions exist. The dimensions are there all the time along with us and our separate consciousnesses and all the other objects and creatures and people we see and interact with when we are looking. We assume this because it's comfortable and because it works. It fits perfectly well with our Cartesian mindset. All this works quite well if you don't examine what's going on too carefully. The moment you begin to examine what's happening in a disciplined way the edges of the experience begin to fray.

R. Buckminster Fuller said: "No one has ever seen outside himself." I go to the window and see the squirrel. He's there. I assume that I see him sitting on the grass, in my yard. I assume that I see him "out there." But, in point of fact, if I examine carefully what's going on, light, coming from the squirrel goes into my eyes after being focused by the lenses in my eyes. It then falls on my retinas forming an image. The retinal cells feel the touch of the photons and send signals to my brain and, while it's true that I see the squirrel it's not true that I see him out there. Instead I see him in my brain. That's what Fuller meant when he said no one has ever seen outside himself. If you understand the mechanics of vision the truth of that statement is intuitively obvious. Everything anyone sees is in his or her brain. Everything that's *out there* is really in our heads. *Out there* is in our heads.

THE MECHANICS OF PERCEPTION

Our perceptual machinery is always on. The channels we rely on to experience the world, vision, hearing, touch, taste and smell, are always open. Even when we're asleep, for some portion of the sleep period, most of us dream. When we do, we access the same experiential channels we use when we're awake. Our minds create dream-space and, in that space, we experience many of the same perceptions we do when we are awake. Some people are able to take control of dream-space through a technique called lucid dreaming. When they do, they're able to do things that can't be done while awake; things like flying, talking to people who are dead, and creating worlds that could not possibly exist while awake.

Let us begin our exploration by trying to bring a more complete understanding to how we experience the world. We have five senses or perceptual channels available to us; vision, touch, taste, hearing and smell. These channels provide access to perceptual domains or realms.

- Vision allows us to access the Visual realm by enabling us to see
- Touch allows us to access the Tactile realm by enabling us to feel
- Taste allows us access to the Chemical realm by enabling us to distinguish between different chemical experiences like sweet, sour, bitter and salty
- Hearing allows us access to the Aural realm by enabling us to detect sound
- Smell allows us access to the olfactory realm because, like taste, it enables us to distinguish between different chemical experiences or smells associated with molecules suspended in air.

We are multicellular beings. Unlike single celled organisms, like an amoeba, our bodies are composed of trillions of cells. All these cells are connected but some of them are highly specialized and function as channels which allow access to the perceptual realms.

Retinal cells in our eyes are sensitive to light. They cannot see but they know when they've been *touched* by a photon that has a specific frequency and wavelength because they can feel that touch. When they are touched they report the experience to our brain.

Tactile cells, throughout our bodies, are sensitive to touch. They cannot feel but they report the experience of *touch* to our brain.

Taste buds, on our tongues and in our throats, are sensitive to specific molecules which produce the sensations of sweet, sour, bitter and salty. When they are *touched* by the appropriately shaped molecule they report the experience to our brain. They cannot taste anything.

Hair cells in our inner ears are sensitive to vibrations. When they are *touched* by vibrations within a certain range they report the experience to our brain. They do not hear anything.

Olfactory cells in our nasal passages are sensitive to specific molecules which produce the sensation of smell. When they are *touched* by the appropriately shaped molecules they report the experience to our brain. They do not smell anything.

These highly specialized cells are the channels. They allow our brain to *tune* the perceptual realms much like a television tuner is able to distinguish between signals broadcast over different frequencies. The seeing, hearing, smelling, tasting and touching are experienced by the organism, as a whole, mediated by channel specific information reported to the brain.

Visual and aural sensation is reducible to touch. Taste, smell and touch are associated with the ability of cells to become aware of external stimuli through changes in potassium ion concentrations on their membranes. In this way they become reducible to the chemical realm or what we, as multicellular organisms, experience as taste. Chemical interaction does not require a cell membrane and, in that sense, predates life itself.

Vision and hearing are reducible to touch. The ability of specialized cells to feel a certain kind of touch is what opens the doors to the visual and aural realms. Those doors are opened when our brains decode the information that the sensory cells send after they've been touched. What these highly specialized cells, in our retinas and inner ears *feel* helps to define and shape reality. Taste, smell and touch are sensations associated with the chemical realm. They add to and help to complete our experience of what we call reality.

We cannot feel space or space-time. We can, however, feel space-time—time or the 0-60 in 5 seconds; per second of acceleration.

> Steady motion has no feeling whatsoever. Gravity on the surface of the earth is a fixed position in a curved space-time. What we feel in our bodies is a constant acceleration upward against the downward "pull" of a curved space-time. Space-time is curved into space-time—time. So we

feel space–time—time even when we are not moving. (If there were no gravity, we would not.)[13]

Let us examine carefully what this means with an eye toward the underlying physics.

Imagine, for a moment, that you're in a big city and you're standing at the corner of a busy intersection. It's sunny and warm and as you look around you see people walking by and traffic moving swiftly on the streets. You look up and see a flock of birds flying, swooping and diving off to your right. You see all these things but you do not feel them. The concept of space and things rushing past you through space comes from vision—from what you see. Space arises from vision.

If you close your eyes and pay attention only to what you feel you'll notice you feel your weight on your feet. Most people assume they're feeling gravity *pulling* them down against the surface of the sidewalk. And while that idea works for most people most of the time, like so many other familiar phenomena we've already examined, it becomes iffy when viewed in terms of how we understand the physics of what's really going on.

To get a better understanding of why what you feel is a constant acceleration upward against the downward pull of a curved space–time let's do another thought experiment.

Now imagine you're in a large well lit room with a table and a chair. There are no windows and you're sitting in the chair looking around to get your bearings. You get up from the chair and notice that, on the far wall, there's a dial labeled gravity. The pointer on the dial is set to 1. The dial is graduated in increments of 1–5 toward the right or clockwise with a small metal post at the bottom, or six o'clock position, which prevents it from being turned any further clockwise. To the left of the number 5, there is a 0.

To the left of the number 1 the numbers, moving counter clockwise run from 0.5 through 0.1 with the final number, after 0.1 being 0, to the left of the little metal post.

When you turn the dial slightly to the right, setting the pointer between 1 and 2, you immediately feel yourself get heavier. When you move it to 2, your bodyweight doubles. You feel much heavier. You try to walk and, while you can, it isn't easy. You move the chair and notice that it's twice as heavy as it was before you moved the dial to 2. You realize that if you continue moving the dial clockwise, it won't be long before it will be difficult to move in any direction.

You move the dial back to 1 and your weight feels normal again. Now you try moving the dial in the opposite direction halfway between 0.5 and 0.4 and you immediately feel yourself become lighter. You walk around a bit and it feels good. You feel stronger. When you move it all the way to 0.3, you're lighter still and walking is even easier. It's also a lot easier to move the chair. You continue to move the dial counterclockwise slowly and, as you do, you feel lighter and lighter. By the time you get the pointer halfway between 0.3 and 0.2, you realize you're having difficulty remaining in contact with the floor and, if you were to move it all the way to 0, you'd begin to float around in the room along with the table and chair. So you move the dial slowly back to 1 and sit back down in the chair.

How could a room like this possibly work?

Imagine that the room is a box in deep space. There is a large rocket engine attached to the underside which you cannot see or hear and, with the pointer set to 1, its accelerating upward producing 1g of downward force on the floor inside the room. When you turn the dial clockwise, the rocket engine provides more upward thrust which pushes you down harder against the floor. When you turn it counterclockwise the thrust decreases and, if you move the pointer to 0 the rocket engine shuts off.

If you didn't know about the rocket engine or that the room was in deep space, with the dial set to 1 you'd *feel* the same as if you were walking around in your house. You'd have that same sense of being pulled down against the floor by gravity. But, in this room, what's *pulling* you down isn't gravity but rather the upward thrust provided by the rocket engine causing the room to accelerate.

Einstein figured this out when he formulated his Principle of Equivalence between gravity and acceleration. Acceleration and gravity *look* very different but they *feel* exactly the same. The *pull* you would feel in such a room, with the dial set to 1, would be indistinguishable from the gravity you feel when you're on the surface of the earth. And that's the reason that what we feel is a constant acceleration *upward* against the downward "pull" of a curved space-time. In physics-speak upward acceleration and gravity are the same thing.

If the only perceptual realm available to you was tactile (touch) meaning you could feel but you couldn't see, smell hear or taste anything, you would not have any concept of being anywhere in particular or of going anywhere. You would not have the perceptual means of understanding what the word space means. There would be no space.

It's difficult to grab onto the concept of being at rest on the surface of the earth while, at the same time, *feeling* acceleration. But, in terms of the physics, the Principle of Equivalence between gravity and acceleration, that's exactly what you do feel. That familiar downward pull is exactly the same thing you would feel in a room accelerating upward in deep space.

On the surface of the earth we're at rest in space-time. If we look around, we don't appear to be going anywhere. What's important to keep in mind, however, is while we're at rest in space-time we are *not* at rest in space-time—time and that's the reason we feel gravity.

Concepts about how the world works are, by definition, enshrouded in labels. We acquire them, file them away and move on with our lives. But, once we begin to examine some of them, in a careful and disciplined way, we discover that, to more fully understand what's happening we need to rethink and expand our perspective. That's what I tried to do with the thought experiment of the room with the gravity dial; to provide an expanded perspective of what's really happening when you feel your weight being pulled down against a chair or the sidewalk. For a physicist, it's intuitively obvious that what he's feeling is a constant acceleration upward against the downward pull of a curved space-time.

But, for most people not familiar with physics or relativity theory the idea is neither intuitive nor obvious.

Everyone knows what the labels up and down mean. If someone asks you what direction is up? You'd raise your arm, index finger extended toward the sky. Few people would argue with that. The problem is, when we point toward the sky we seldom take into account that we're standing on the surface of a very large sphere and most of the time, because the sphere is so large, relative to us, the world looks like it's flat.

Now, imagine that you are standing on the North Pole and your best friend is standing on the South Pole. I'm in my living room in California. I pick up my cell phone and establish a conference call between you, your best friend and me. Then I ask you both, which direction is up? Since all three of us are on the surface of a very large sphere you and your best friend are pointing in opposite directions relative to me. If one of you were to ask me which direction is up I'd be pointing in a direction about 90 degrees different from the one both of you are pointing relative to the surface of the earth. The concepts of up and down are relative concepts. They have no meaning in deep space.

A more scientifically accurate description of the direction we are all pointing when we think we're pointing up is *out*. We're all pointing in a direction that is out or away from the surface and the center of the sphere we're standing on. Perspective changes everything. Pilots know this intuitively. When they talk about landing they say "I'm coming *in* for a landing."

The labels we choose and use shape the middle latitudes of reality. People often talk about watching the sunset or watching the sunrise. It's romantic and poetic but it's wrong. Relative to us, on the surface of the planet, the sun doesn't move at all. It neither rises nor sets. It just sits there 93 million miles away. What we're really seeing, when we watch a sunset or a sunrise is the earth spinning on its axis. The movement of the earth on its axis makes it *look like* the sun is rising and setting but it's not. The movement looks slow but that's just because the earth is so large. It's not slow. It's fast.

Examining labels carefully, with discipline, often expands understanding by changing perspective.

Vision and touch are linked by the photon because a photon can be both felt and seen.

> It is both felt and seen. It is the smallest unit of visual consciousness, and a single point of energy in space, time, and mass. Nothing smaller can be seen, no matter how powerful the microscope, because the space in which it is seen does not get any smaller. The photon produces the space that it is in. It is like a dot in a newspaper photograph. The picture arises from a spatial arrangement of many dots, and can be no smaller than a single dot. But as a unit of energy, a single photon can be felt as well as seen. It can be either a particle or a wave. Individual cells "feel" individual photons. That, in fact, is what I think a photon is. It is the experience of an individual retinal cell as reported to the organism as a whole. The cell feels, the organism sees.[14]

It is at this point that we begin to get some idea of where Samuel Avery is going as he reveals his understanding of how vision works. Vision creates space. Photons create vision. Therefore, visible photons are visual consciousness.

An individual retinal cell never *sees* anything. It has a life of its own that we know nothing about. When touched by a photon with a specific wavelength and frequency it reports that experience to your brain. Your brain/body (your entire organism) interprets that message as a dot. The color of that dot depends on the frequency and wavelength of the photon that touched the retinal cell. That dot is space. Trillions of photons bombarding millions of retinal cells produce vision. Space arises *from* vision. Light *is* visual consciousness.

If this is true, and I believe it is, it follows that space–time is in light and not the other way around. We normally think about light as being in space–time but if light is visual consciousness and space arises from vision, space–time is in light.

Individual retinal cells or, for that matter, any kind of individual cells know nothing of space because while they can touch and taste they cannot see or smell or hear. The same would be true for you or me. If all we could do was touch and taste space would not exist. For individual cells and multicellular organisms like plants, space does not exist.

It is extremely difficult if not impossible to imagine what it would be like if all we could do is feel the sensation of touch without being able to see, smell, hear and taste. And, if the only perceptual domain or realm we had access to was touch our experience of touch would be very different than it is when it is coordinated with vision, smell, taste and sound. We are wired to receive continual stimulation from all five perceptual channels; so much so that when we are purposely subjected to sensory deprivation, before long, we begin to hallucinate. One reason that's so is, sensory deprivation makes space and the world go away.

When a photon touches one of our retinal cells the cell reports that it's been touched. But, that's all it does. It doesn't see anything. We, on the other hand, experience that report as a dot of space–time. One photon is equivalent to one unit of space–time which is the smallest unit of space–time we can perceive. But the photon is special because it exists simultaneously in five separate dimensions; three space dimensions normally associated with the xyz Cartesian coordinates, one time dimension and one frequency or (per time) or (time-time) dimension. Because this is so, the photon is able to relay information in five completely separate, but nonetheless coordinated domains or realms of perceptual experience.

Since the time of Descartes, we have learned much about the world and how it works. We know things Descartes couldn't possibly have known when he created the prevailing myth which supports the existence and primacy of matter. Descartes knew nothing about the existence of quarks or atomic structure. Quantum mechanics was yet to be discovered. The special and general theories of relativity were unheard of. In the time of Descartes scientists believed the universe was static and that time flowed smoothly regardless of how fast someone traveled. No one had ever heard of or even imagined the existence of a black hole, a neutron star or a quasar. There was a place for everything in the universe and everything was assumed to be in the place it belonged.

But now, that we know space and time curve and blend into one another; that, as we accelerate time slows down for us relative to someone who is standing still; that, at the event horizon of a black hole, time comes to a screeching halt; that light behaves as both a particle and a wave but never as both at the same time; that subatomic particles of matter behave the same way that photons do when subjected to the classic double slit experiment; that, in terms of the physics, gravity and acceleration are indistinguishable and that, on the quantum level we are forever forbidden from knowing both the momentum and position of subatomic particles; this new and expanded knowledge is beginning to tear the fabric of the myth Descartes created.

We can choose to ignore what we know but we do so at our own peril. Given what we now know the fabric of Descartes' myth is frayed, tattered and torn. The intellectual foundation it once represented which we relied on to help us understand how the universe works is crumbling. When it finally falls away we will need a new myth; one that incorporates what we now know.

Old myths are easy to cling to because they are comfortable. Once they lose their explanatory power, however, comfort must be exchanged for new myth which better explains how the universe works. Believing that we existed at the center of the universe and that all the celestial bodies moved around the earth was a comfortable myth that became entangled with religion and lasted for centuries. Eventually we came to understand that it was our planet that was moving relative to the sun. The geocentric myth fell. We now understand that the earth moves and is not flat.

The prevailing myth regarding the existence of matter and of the universe as a container with objects and people in it is, in light of what we have discovered, shaky at best.

The chittering sound in my back yard is not a squirrel. The bark I hear in my neighbor's yard is not a dog. The droning sound I hear in the sky is not an airplane. These things are sounds that exist in one of the five perceptual realms. The path to the new myth is in the direction of bringing more, expanded and clearer understanding to how these sounds are coordinated with the other domains or realms of perception. It is also in the direction of bringing expanded understanding to consciousness and what the word consciousness means relative to whom and what we are. The prevailing Cartesian myth posits that each of us *has* consciousness. A new myth may force us to abandon that belief.

The Cartesian myth posits the existence of matter and defines reality in terms of its existence. Things other than matter—things of the senses like taste, color and smell are not to be trusted. A new myth will reject the existence of matter and expand the definition of reality to include not only taste, color and smell but also other nonobjective experiences like dreams, emotions, spirits, visions and ideas.

Science, the scientific method, and technology have enabled us to create a world grounded in objectivity and expanded our understanding of how the universe works. Scientific discipline insists on objectivity and rejects experiences associated with subjective domains. But in exploring phenomena like the dual nature of light, black holes, quantum mechanics and dark matter, it has begun to probe aspects of the universe that fail to conform to the rigid standards prohibiting inclusion of subjective aspects of existence. We now know that certain scientific experiments require the presence of an observer and that the presence of an observer affects the outcome of those experiments. Like it or not, subjectivity has forced its way into the heretofore iron clad rigor of carefully controlled objective scientific experiments. Previously, subjective and objective were separate and unequal. Now, they have become inextricably entangled which forces us to seriously examine the possibility that, instead of being separate and unequal, they might be opposite sides of the same coin.

Let us explore what we know about how the universe works using knowledge gleaned from disciplined scientific investigation and see if, in doing that, we might find the building blocks and mortar required for the foundation of a new myth to replace the one that's beginning to crumble.

Experiences we have in any or all of the perceptual domains or realms are objective because they can be verified by other observers. Imagine that you and I are sitting across from each other at a table. There is a glass sitting on the table filled with a colorless fluid that is effervescent. When I ask you if you see the glass you tell me you do. You also confirm that, like me, you see bubbles rising in the fluid. I lift the glass and bring it to my nose and, while I can feel the bubbles bursting on the surface of the liquid, on my nose and lips, I tell you the fluid has no smell and hand you the glass. When you bring it to your nose, your experience matches mine and we agree the fluid has no smell. You then take a sip from the glass and put it down. You tell me that it's cold and tastes like seltzer. Then you bring the glass to your ear and tell me that you can hear it fizzing after which you hand the glass to me. I do the same things you did and confirm that the fluid is cold and tastes like seltzer and, after bringing the glass to my ear I tell you that I too can hear it fizzing. We were able to perform all this verification using all five sensory realms.

This is a demonstration of objectivity. We both see the glass and touch it where we see it. We also smell, taste and hear what's in the glass where we see it. And, finally, we agree that the glass exists and is filled with a cold, colorless, odorless and tasteless effervescent fluid which we identify as seltzer. The verifiability of perceptual experiences like the ones described above is what makes them objective. If there were five other people sitting at the table with you and me they would all verify what you and I experienced when we examined the glass and its contents.

Exchanging the Box for the Screen

R. Buckminster Fuller defined universe for each individual as: "everything that isn't me plus me." Viewed from that perspective it's easy to think of the universe as a huge container or box. I am inside that box along with all the other people and things I see or encounter. This is a fairly trusty model. For the most part, it holds up well and helps us to bring understanding to what we observe going on in the box. Using this model we've discovered physical laws which govern the way objects interact with each other and with us. The universe as container or box makes sense until we ask; what is the box in? Where is the box?

Fuller said, "It is impossible to escape from universe and realize that escape has occurred." This is an interesting concept but it doesn't help to explain where the box is. If I could escape from the box, where would I be? And, here's another question. If it's true that it's impossible to realize that escape has occurred, does that mean I would not be able to see the universe I had escaped from; if so why not?

As we continue to examine the box we notice that other people have experiences similar to our own. They see things where we see them. They also smell, hear, taste and touch objects where, like you and me, they see them. We look at the glass on the table and assume that photons are bouncing off it and into our eyes. Then those photons are focused by the lenses in our eyes onto our retinas which send signals to our brains and somehow those signals are converted into an image of the *real* glass sitting *out there* on the table. Then we take our assumption further to say what we see is only a representation of the *real* glass and to explain that, we assume the reason we see the glass is that we all have this magical substance called consciousness. We don't know what it is but we're pretty sure it's somehow associated with our brains. When we begin to probe the question of exactly what consciousness is, we run into problems and questions that nobody seems to be able to solve or answer.

Zion's axiom demonstrates the extent to which we don't understand consciousness or being. Consciousness or being somehow arises in the brain and everyone has it.

It seems that consciousness is necessary to explain what's going on in the world around us. It also seems that in all this, consciousness is the only true unknowable. So let's make some assumptions about consciousness before we proceed with exchanging the box model of the universe with the screen model.

- Consciousness is the only unknowable
- Consciousness and being are one and the same
- Consciousness cannot be located in the brain or in electrons or in the firing of neurons in the brain or in the exchange of neurotransmitters like acetylcholine between neurons in the brain

In *Transcendence of the Western Mind*, Samuel Avery replaces the box with the screen in order to put doing inside of being.

Before Einstein we believed space and time were absolute and fundamental. But, with special relativity we have learned that they are neither absolute nor fundamental. Space and time blend and curve into each other and form a kind of fabric that can be warped by objects that are massive.

When we examine matter very carefully we find it is impossible to scientifically prove that it exists. That it appears to exist at all seems to be inextricably entangled with consciousness. Let us make another assumption—Consciousness Is Everything!

Perceptual consciousness is a multidimensional screen. Objects that we perceive using vision, hearing, touch, smell, or taste or using some combination of those perceptual realms are located on the screen. Space and time provide the structural framework of consciousness that coordinates and organizes what we see, hear, smell, taste and touch on the screen. If there is nothing on the screen it is blank and space-time is empty. Empty space-time is potential perception.

What we perceive as matter, using one or more of the perceptual realms available, is really the coordination of those separate realms of consciousness into existing dimensions.

Dimensions correspond to and are coordinated with the perceptual realms or domains. Here's what I mean. If I see an object, like a glass sitting on a table, I can touch that glass because the tactile and visual realms or domains of consciousness are perfectly coordinated at that particular point in both space and time.

Now, to truly grasp the difference between the box/container model of the universe versus the screen model, we need to reorient our thinking about what happens when we choose either model.

If we choose the familiar box/container model, we think about light and sound bouncing off material objects.

If we choose the screen model we need to abandon that idea. Instead, with the screen model, light and sound do not bounce off anything. Light *is* the visual realm or domain of consciousness. Put differently light *is* visual consciousness.

And so it is with sound. Sound *is* the auditory realm or domain of consciousness. Sound *is* auditory consciousness.

This is a significant change in the way we think about what's going on in terms of how we experience the universe. But it is important in bringing understanding to what Samuel Avery describes as The Dimensional Structure of Consciousness. This dimensional structure or framework is the linchpin which holds the screen model together.

As I continue to describe how the screen model works and why it's better than the box/container model, if you choose to come along for the intellectual ride, you will be forced to reorient your thinking in ways that, I promise, will be painful and difficult. Not so much because they are complex, convoluted or intricate but because they're so different from your normal ways of thinking about what's going on that you will be sorely tempted to chuck it all and stick with what feels comfortable. This will be especially true if you have had formal training in one or more of the hard sciences like physics. You'll be faced with the same predicament that faced humanity when Copernicus realized the earth was not the center of the universe. At the time, when he suggested it was the earth that was moving rather than the sun, the initial reaction was *that's crazy*. If it's true that the earth is moving why don't I fly off its surface? Why don't all the mountains and buildings crumble and fly off the surface? Why does the sun rise and set every day? The earth is flat and continues to be flat past the horizon. Copernicus must be nuts. Myths and beliefs die hard.

If you choose to come along for the ride, remember that, when it's over, you can choose to conclude that Samuel Avery and I are nuts and stick with what you already know and feel comfortable with. My goal here is not to force anyone to change their thinking about how the universe works but to offer alternative explanations for what we experience, how we experience it and why, when considered carefully, the screen model works better than the box/container model. Think of it as a thought experiment and let the chips fall where they may.

In exploring how the sensory realms work and examining their roles in how and why we experience the universe we return, repeatedly, to what seems to be both an intellectual as well as an operational coda—*Light*.

If we think of all five perceptual realms as a screen on which objects we perceive are located, we find that the fundamental structure of that screen is derived from the structure of light rather than say, sound or any other perceptual realm. The auditory, tactile, olfactory and chemical realms are experienced at the same place and time where they *would be* seen.

Instead of experiencing five separate one-dimensional worlds we experience a single five dimensional world because our mind uses the dimensional order in which we see things as a format within which we experience the non-visual realms. Put differently, that format informs me that when I feel, smell, taste and hear an object, that is where I *would* see it if I looked. If I touch the underside of my chair I experience its surface using the tactile realm and, I know that if I get down on the floor and look, I will see what I felt exactly where I felt it.

Sound or smell could have been used as the format for the dimensional order in which we experience objects on the screen instead of light but light, unlike sound, smell, touch or taste, is so close to being a perfect format for all of perception that we didn't even know about its imperfections until the twentieth century.

Attempting to envision the universe as a screen instead of a box is difficult because we're invested in the box. We're comfortable with the idea that everything real is in the box and that everything that's in the box has a place in the box. But, as I pointed out earlier, it is impossible to imagine where the box is. There is no satisfactory answer to the question; where is the box?

Absent being able to get some idea of where the box is it becomes impossible to find or establish any kind of meaningful context, within which, we can begin to understand it. The model works well enough if we don't ask too many questions and if we don't focus our questions too tightly.

If we think of visual consciousness as a four dimensional screen we're faced, once again with the question; where is it or what is it in? And, at this juncture, for visual consciousness or—*light*—the screen itself provides the concept of inness. It cannot be a context for itself because the only appropriate context for the screen is the whole of consciousness.

Here is where the magnitude of the challenge begins to unfold regarding substituting the screen for the box. To be successful you must be willing to abandon the box completely. It's one thing to say that but something altogether different to do it. The reason it's so difficult is because the Western mind set based on Cartesian principles and all the science associated with those principles is solidly invested in the box. Einstein, arguably the greatest scientific mind of the twentieth century, was so strongly invested in the box, he refused to accept the fundamental principles of quantum mechanics. His famous retort to Heisenberg—God does not play dice—captured how strongly he was invested in believing that, ultimately, everything about how the universe is structured and works could be predicted with a high degree of accuracy. He could see the value in the discoveries his colleagues were making in the realm of quantum mechanics, but he could not accept that, at the quantum level, uncertainty reigned supreme. After general relativity, he spent the rest of his life attempting to tie together the principles of gravity and electromagnetism to produce a theory of everything which he believed, if he could find it, would nullify all the uncertainty associated with quantum mechanics. He died trying to do that and, for all his brilliance, he was wrong. He showed the world that space and time curve and blend into each other, that light always travels at the same speed and that it behaves as both a particle and a wave, that gravity exists because massive objects warp space-time and that energy and matter are opposite sides of the same coin. The tenacity with which he held onto the Cartesian and Newtonian views of how the universe works disallowed him from being able to embrace the abject weirdness associated with what his colleagues were discovering about the quantum level of reality.

Individuals who are not so heavily invested in Western thinking and the views of Newton and Descartes may find it easier to explore the screen model because they are not as profoundly invested, intellectually, in the box. Nonmaterial aspects of existence are common amongst people with an Eastern mind set.

Nondimensional experience of the kind gained through the practice of meditation dilutes the strength of the box/container paradigm.

Western thinking is the foundation of physics as we understand it today. It is Cartesian, Newtonian and box oriented. But the central problem seems to be that it has outgrown the metaphysics in which it is so heavily invested. In deeply probing what happens on the quantum level of reality, the hardest of the hard sciences is faced with enigmas that, under normal circumstances, appear almost mystical. The behavior of photons and subatomic particles has been experimentally proven to *depend* on the presence of an observer. Subatomic particles can be in two different places simultaneously and if two particles are purposely entangled, making an observation regarding one of those particles has an instantaneous affect on the other particle, regardless of the distance between the two. Such things are not supposed to happen in the box.

Part of the ongoing challenge for modern physics is to reexamine its metaphysical underpinnings. In doing so it may discover that light can be better understood using the screen model rather than the box. The box model searches for light in space–time and runs headlong into enigmas which appear mystical. The screen model, instead, finds space–time in light. Asking Western physicists to cross that intellectual chasm is indeed a tall order but I believe it points the way toward a metaphysic that will help explain what the box model cannot and bring deeper understanding to the physical phenomenon of light.

> "Light, as visual consciousness, is a form of being. If we come to understand dimensions within light, and doing within dimensions, we have come to understand doing within being."[15]

Exploring the screen

The screen which represents the entirety of perceptual consciousness is the quantum screen. The photon screen is part of it but the quantum screen is not limited to visual consciousness alone. It extends beyond vision in every direction to include the auditory, tactile, olfactory and chemical domains of consciousness.

When you touch the back of your head you are experiencing the back of your head on the quantum screen. You can feel it via the tactile realm of consciousness and you know that if you went into the bathroom which has two mirrors you would be able to feel and see the back of your head. In the bathroom, with the mirrors, you would be experiencing the back of your head in both the tactile and visual realms of consciousness. But, without the mirrors, the photon screen is unavailable and you must rely on touch or the tactile realm; (quantum screen only—minus the photon screen) to explore the back of your head.

See the container of French fries on the table. They're hot and you can smell them and see salt crystals clinging to the surface of the fries. You're experiencing the fries on the quantum screen plus the photon screen. Pick up a fry and put it in your mouth. The moment you do that you can feel it on your tongue and against your teeth. You can taste the salt and the flavor produced by the smell of the potato and the oil in which the fry was cooked. You hear the sound of your teeth as they crunch through the fry's crispy surface. All these sensations are quantum screen minus photon screen. That is essentially how the screen model works.

Where you see something is where you can *potentially* touch, taste, smell or hear it. Hearing and smell are not as clearly defined as touch and taste but in terms of bringing understanding to how the screen works, in general terms, where you hear or smell something is where you would see it.

Seeing an object, unless you've seen it before, tells you nothing about how it will taste, feel, smell or sound. We explored this idea earlier when examining mass and boxes filled with lead shot.

Seeing a plate of onion rings sitting on the table tells you nothing about how they will taste. They could be hot or cold, salty or sweet, bitter or sour or they could be made of paper, rubber or plastic and painted to look like the genuine item. When you see the onion rings the only reliable information you have is where you need to go if you want to taste them. This is a subtle but important aspect of how the screen model works. When you *see* the onion rings all you really know is, if you go to where they are, the *potential* exists for you to taste them, touch them and smell them. The point of all this is, what applies to the onion rings, applies equally to the entirety of the concept of material substance. It doesn't matter what kind of object(s) we might be talking about. Instead of onion rings we could be considering a glass, or an elephant or a bird or an ice cream cone. What's important to understand is the entire concept of material substance can be understood as *potential* perception.

This separation of information exists for and pertains to all five realms of consciousness. Touching an object provides no information regarding how it will look, smell, sound or taste. Tasting an object tells you nothing about how it looks sounds, smells or feels. Smelling an object may provide limited information about how it might taste in terms of flavor but, flavor is associated with smell more than taste meaning, once again, unless you've smelled and tasted the object before, you'll have no information regarding whether you can expect it to be salty, sweet, bitter or sour. Hearing the sound an object makes provides no information regarding what it will look like, taste like, smell like or feel like. All the relevant information about an object is on the screen but it must be accessed through the appropriate combination of realms or domains of consciousness.

From the standpoint of your own experience, by now, you should have some appreciation that physical reality exists without the need of substance or matter. The existence of matter cannot be scientifically proven. The reason you touch something where you see it is explained by the dimensional structure of consciousness. I know what you're thinking. Why do my friends and other people see the same truck, glass, car, girl or building etc. that I see?

It is here, at this juncture between the box and the screen, that I will ask you to abandon the idea that your friends and other people *have* consciousness in their brains or anywhere else in their bodies. Forget about that and focus all your attention only on what you *actually* experience.

For the same reason(s) you do not have any direct experience with matter, you do not have any direct experience with the consciousness of other individuals. You assume that matter exists in objects and consciousness exists in other individuals to help you make sense of what's happening in the box. But if you exchange the screen for the box you can abandon the idea that others have consciousness, in the same way and for the same reasons that you can abandon the existence of substance or matter in objects.

That other individuals *appear* to have consciousness, is explained by the same structure of consciousness that causes objects to *appear* that they are composed of matter or substance.

If you've come along for the ride so far, I'm going to ask you to abandon one more thing; the idea that *you* have consciousness.

The screen model works only if you assume that nobody *has* consciousness. Assume, instead, that everyone and everything is *in* consciousness. Self is *in* consciousness along with everything else.

You, other observers that you see, family members, lovers, close friends and enemies have neither consciousness nor matter in themselves. We've been taught to associate consciousness with self; to believe there's an inextricable link between the two. The key premise of the screen model is that *consciousness is independent of self.* The self exists so that it may do but not so that it may be.

Objects on the screen fall broadly into two categories, animate and inanimate. Inanimate objects, like chairs and rocks, cannot move of their own volition. They have no volition. Animate objects like people, dogs, cats, squirrels, algae, bacteria, plants and insects can and do move of their own volition and, for the most part, they move unpredictably. They all have access to one or more realms of consciousness. They are alive. They are observers.

Observers are an order of consciousness higher than self. The only way to realize that this is true is through transcendence of self. Consciousness or being is not in self nor is it in the doing that creates self. The answer to the question (*why are we here?*) is *we are here to observe.* These propositions are the foundation for a new and different metaphysic which better explains how and why we experience the universe in the ways that we do.

The self can be transcended through the practice of meditation.

Observers and the Observational Realm or Domain of Consciousness

What distinguishes observers from inanimate objects? We've already established that they are alive. Inanimate objects don't move unless someone moves them. Observers move of their own volition. They can also navigate around inanimate objects and other observers. Their motion appears to be orderly and observers we identify as people along with, to lesser extent animals, are able to communicate with each other about what they are experiencing.

When another observer tells you about something he experienced, like having seen a squirrel in his backyard, you don't actually perceive the squirrel. If he's telling you about the squirrel all you perceive are him and the sounds coming from his mouth. If he's telling you about having seen the squirrel in a letter or an email, all you perceive are the letters, punctuation and words in his missive.

What's important about these sounds and letters, punctuation marks and words, is they possess an intrinsic order which makes them more significant than ordinary sights and sounds. We recognize that perceived order as language. Beyond the communicative and literary properties associated with language it has physical properties. The order embedded in language is much like the order you notice when you see how observers who are not communicating with each other are able to keep from bumping into objects that happen to be in their way. The key difference is the order inherent in language is of a much more highly refined form. In thinking about this it's important not to confuse the parts with the whole. Random sounds, words, letters and punctuation marks are not language for the same reason that a pile of bricks and some mortar is not a building. But it's also important to recognize that a building is, *ultimately*, reducible to bricks and mortar in the same way that language is, *ultimately*, reducible to sounds, words, letters and punctuation marks.

The individual parts have extremely limited significance but they can be and are used to construct objects and information rich with meaning and infinitely variable.

With language what we actually perceive is limited to sounds which are akin to noise and visual objects which appear on paper, as ink spots or on an electronic display as the equivalent of ink spots. It is the *order* in which these sounds and ink spots are arranged which provides access to a sixth realm of consciousness—the observational realm or domain of consciousness.

Understanding how all this works requires becoming and remaining mindful about how the parts relate to the whole.

> "Observational consciousness is an orderly construction of perceptual parts the way visual consciousness is an orderly construction of tactile parts."[16]

In the same sense that visual consciousness is reducible to the tactile sensations retinal cells experience when they are touched by photons, observational consciousness is reducible to the sounds we hear and the ink spots we see associated with language.

Much like the perceptual realms, the observational realm of consciousness arises when triggered by appropriate circumstances or stimuli. Space–time arises when vision is triggered by the touch of light on retinal cells and becomes a wholeness above and beyond vision. The observational realm arises when perceived language provides the orderly stimulus required to interpret what is seen or heard. When the observational realm arises it becomes a wholeness above and beyond the sights and sounds, of which, it is constructed. It is created by orderly arrangements in the perceptual realms. But, beyond that, it is important to understand that the moment you accept the validity of what someone says or writes it becomes as real as if you had perceived it in one or more of the perceptual realms. In this sense it becomes *potential* perceptual consciousness because you would see, hear, smell, taste or feel whatever an observer tells you he sees, hears, smells, tastes or feels if you were where he is.

Here's what I mean. Imagine that you and a friend are in a house and you are in separate rooms. You can't see each other but you can hear each other if either one of you speaks. Your friend asks, "Why is there a handgun on the coffee table?" If you come into the room where he is and you don't see the gun there is no dimensional coordination and no observational realm. If you do see the gun the observational realm exists (arises) because dimensional coordination exists—you see what he saw when you enter the room he is in.

The observational realm of consciousness is coordinated with the perceptual realms of consciousness in the same way the perceptual realms are coordinated with the dimensions and each other. This is the key to understanding dimensional coordination and the dimensional structure of consciousness.

In thinking about and trying to understand the concepts presented here, you will find yourself reflexively putting consciousness back in yourself and other observers. As difficult as it may be, I'm going to ask you not to do that; to remain vigilant of the reflexive need to do that. After all, most of us have been taught to assume we all *have* consciousness. But this new paradigm is based on the assumption that nobody *has* consciousness; that as an alternative everybody, instead of *having* consciousness, is *in* consciousness.

The value in reversing your assumption regarding where consciousness resides and who has it is that it allows for an understanding which, initially, seems more complicated but, ultimately, becomes simpler.

Dimensional correspondence

Absent the idea of perceptual realms or domains which are associated with or correspond with spatial and temporal dimensions, the box universe makes sense. What it does not and cannot explain are space–time, matter and consciousness in observers. It makes assumptions about these things; that observers have consciousness; that matter exists and that space–time is constant and the same regardless of whether one is considering reality on the level of the cosmos, the macroscopic level where we exist, or the quantum level which underlies cosmic and macroscopic reality. But it is important to remain mindful that these are assumptions because we do not truly *know* anything scientifically substantial regarding space–time, matter and consciousness in observers. So we rely on speculative assumptions and move forward from there.

Dimensional correspondence is a guess based on a set of observable assumptions that dispose of space–time existing in and being the same on all levels of reality, the existence of matter, and consciousness in observers. Instead it postulates that each perceptual realm of consciousness corresponds to a spatial or temporal dimension.

There are five perceptual realms:

- Vision
- Hearing
- Touch
- Taste
- Smell

And there are five dimensions:

- Length
- Width
- Height
- Time
- Mass

Mass appears foreshortened in space–time as a second time dimension or the second per second we normally understand when we say a car accelerates from 0 to 60 miles per hour in 5 seconds. This is the same kind of foreshortening we experience when, while watching television, we see a depth dimension or z-axis on the screen when, in fact, the only physical dimensions the screen truly has are length and width, which, in physics-speak are x and y-axes.

Dimensional correspondence rejects the existence of matter and consciousness in observers. It also proposes the existence of a sixth realm of consciousness, the observational realm, which is not quantitative, less defined and foreshortened in space–time. We recognize it when we see other observers avoiding bumping into objects or other observers in their way. In physics-speak; the *foreshortening* we are noticing is non-uniform acceleration. We also notice that only observers are capable of self directed and purposeful orderly motion.

SWAPPING DIMENSIONS

We experience dimensions all the time but we don't think much about it. Physicists think and talk about dimensions and they like to talk about swapping or interchanging dimensions to describe things we do every day. For most of us physicists' descriptions of dimensional interchange wouldn't make much sense unless related to experiences we are familiar with. Here's an example. A simple space-for-space interchange occurs when we move our eyes. If we're looking straight ahead and move our eyes to the right we've exchanged one space dimension, along the z axis for a different space dimension along the x axis. In physics-speak we've rotated one space dimension into another space dimension. This is a simple space-for-space swap or interchange.

If you get up from your chair and walk across the room in physics-speak you are rotating a space axis into the time dimension. When you swap or interchange dimensions like this, objects move past you and perceptual consciousness appears to move through space.

When you accelerate you execute a more complicated interchange because you're rotating space–time into the mass dimension. When this happens you appear to be accelerating through space.

If you move at a constant velocity, like 60 miles per hour, only one time dimension is involved and, as pointed out earlier, you can't feel constant velocity.

However, if you move from 0 to 60 miles per hour that's linear acceleration and involves two time dimensions expressed as the second—per second or the 0 to 60 portion in a certain number of seconds. This second time dimension is the mass dimension and is another example of dimensional correspondence in that it corresponds to the tactile realm of consciousness because you can *feel* it. What you're experiencing is a g force. You experience a g force when you step on the accelerator pedal, go over a speed bump, hit a pothole, go around a curve or find yourself in a gravitational field.

Initially, all this talk about swapping or interchanging dimensions seems convoluted and complex because most people never bother to examine what's going on the way a physicist would. But what I'm describing is simple. It represents things you do every day like walking down the hall, turning your head or picking up your coffee cup.

Most people associate acceleration with driving or riding in some kind of vehicle like a car. But, acceleration is much more than just that. When you experience acceleration in a car or when riding a roller coaster, your entire body is involved. You feel it throughout your whole body. But, every time you pick up a coffee cup or paper clip you're also experiencing acceleration but only in one part of your body and, more often than not, it's counterbalanced by what you feel in another part of your body so there's no apparent acceleration. You see the cup or paperclip move but you don't think of it as acceleration even though it is.

The self exists in order to do and "doing" *requires* dimensional interchange or swapping. Understanding dimensional interchange enhances and expands understanding of "doing."

You do not need your body to think, dream or meditate but, without your body you cannot "do" anything.

POSSIBILITY OR POTENTIAL

Possibility or potential provides a consequential framework or context for information. Here's an example. By itself, the word charge is practically devoid of meaning. If you put it in a context like, "how much will you charge me to fix my computer?" the potential becomes *how much is it going to cost me?* The situation is (your computer isn't working) and the charge could be anything from nothing, because the person you're talking to is a friend; to several hundred dollars because you had to call in a technician on a Saturday.

It is potential which creates a context for information and makes experience meaningful and exciting. If Jim receives a letter from Mary signed, Love, Mary; if Mary is a colleague with whom he works but to whom he has no emotional ties, the closing salutation Love, is nothing more than pleasant and sweet. If, however, Mary is someone Jim would like to become emotionally involved with, if she sends a letter signed Love, Mary; the closing salutation immediately takes on significantly expanded meaning. The *potential* meaning of the word Love expands appreciably because it might signal that Mary feels the same about Jim as he does about her.

If you place a large bet on a game of roulette you have the potential of winning a large amount of money if the little ball lands on the number you chose. What creates the context and makes this so exciting and interesting is the little ball could land anywhere on the wheel. We create the potential by building casinos and paying people to operate the roulette wheels, craps tables, black jack tables and slot machines. When we do that we create a meaningful context for the *information* surrounding the games and contained in the phrase "we have a winner." Absent that focused context—where the little ball lands on the wheel, how the dice fall, the order in which the cards are turned and the phrase, "we have a winner," mean nothing.

To be meaningful information needs a potential in which anything could happen besides what actually does happen. The value of your 401K portfolio is a potential because it could increase or decrease. That's why you pay attention to it. A page in a book is a potential because the words, letters, and punctuation marks that appear on it could say anything else besides what they are saying. That is how what the author has to say is made available to you. It is potentials that make actual experiences seem *real*.

This applies equally to dimensions. A dimension is a potential because it provides a context or framework for the information associated with perception. Random motion of air molecules or photons, absent the context provided within a dimension, would not *mean* anything—would not constitute or mean physical objects of any kind. Photons and moving air molecules become objects because and only when they are in dimensions. And here again we notice that experiences we have via the senses are interesting and meaningful because, regardless of the experience we are having, it *could be* something entirely different. The steak was perfectly done and delicious—it was overdone and tough as shoe leather—it was underdone, raw and cold.

The five perceptual realms correspond to the five sensory potentials and are integrated into and coordinated with what we identify as space-time-mass or *the screen*. Keep in mind that space-time-mass makes reference to five dimensions: space (length, width, height) which is three dimensions, time which is one dimension and mass which is one dimension. The *potential* for sensory perception exists at any point in any dimension. This is the reason that the whole concept of material substance can be understood as *potential* perception. All actual and potential action and experience happens on the screen.

Experience becomes complex, interesting and meaningful because of potentials. But there is a yin and yang or tensional aspect to what potentials bring to experience because while they bring complexity, interest and meaning they simultaneously bring a sense of nonbeing. Here's what I mean.

If you're in the house and you hear your friends outside laughing and splashing around in the backyard swimming pool you will be struck by a sense of not being with them; not experiencing what they are. If you walk past a bakery and smell the aroma of fresh bread baking, you will have a sense of not tasting it. There is a sense that something is happening in the world but you're not experiencing it.

In the observational realm, the sense of nonbeing is much more intense than it is in the perceptual realms because, by definition, the observational realm is *potential* perception.

People tell you about events happening in your neighborhood or around the world that you do not perceive directly and that creates the feeling that what you are experiencing represents only a small piece of what is *actually* happening. This creates tension which may lead you to believe you are nothing more than a tiny spark of consciousness in a very big box. If potential is a double sided coin then the yin side is labeled excitement, interest, being and meaning and the yang side is labeled nonbeing. The coin of the domain in the dimensional structure of consciousness is potential. It creates tension by giving and, simultaneously taking away a sense of being. It is what we use to purchase the experience we call life.

When someone tells you about what is happening in a place different from where you happen to be, if you think about how the screen model works, you will notice that your actual experience of what you're being told is limited to the words you hear or read. What someone says he or she experiences is, for you, potential only.

What you experience directly through the sensory realms, defines the limit of the extent to which you *actually* experience the universe.

As you think about potential perception and what it is and means you may be tempted to conclude that it isn't real. It is real. It is, in fact, rapidly becoming a much larger and exponentially expanding portion of the whole of consciousness. Language and writing opened the gateway to observational consciousness for observers. Initially, observers needed to be in close proximity to each other to experience observational consciousness through language and to exchange messages and information written down on clay tablets or papyrus. Over time, technology expanded the reach of observational consciousness with the introduction of things like the printing press, the telegraph, the telephone, radio and television. Today, we are experiencing a much more rapid expansion of observational consciousness happening so fast it is a bona fide spike compared to earlier expansions and the timelines associated with them.

This spike in the expansion of observational consciousness is being facilitated by the availability of ever expanding amounts of computational power at ever decreasing cost to ever more people than was ever possible. Combined with the existence of the internet and cheap cellular phone service we are watching and participating in a spectacular spike in the expansion of observational consciousness by using the new technology to create venues which not only facilitate but, in a very real sense—force—exponential expansion of the observational realm. Some examples are websites like Digg.com and Google. Others include websites which allow users to create web logs or blogs and the ones I believe are having the greatest impact are represented by websites like My Space, Facebook and Twitter. Of these three, Twitter is exerting the most disruptive expansion of observational consciousness because it was designed, from the outset, to facilitate communication in real-time. Users exchange information or tweets about what's happening around them with the ease they would have if they were face to face. But Twitter amplifies the experience by making it instantaneously available to *every* user who happens to be logged on to the service and participating. This computational web based amplification is what's powering the spike in expansion.

Here I'm reminded of what R. Buckminster Fuller called ephemeralization which he defined as our ever expanding ability, using technology, to do more and more with less and less. The example he used, in the 70s was comparison of the communications capabilities afforded by the first transatlantic cable versus the first Telstar satellite.

The first transatlantic cable, laid in 1858, was 1600 nautical miles long, weighed 171,200 pounds and could send Morse code at the rate of 0.1 words per minute or two minutes per character.

Telstar I launched in July of 1962 weighed 170 pounds, was about 35 inches in diameter and could handle 600 one-way voice circuits or one TV channel. Or it could handle 60 two-way voice circuits simultaneously, limited by ground equipment.

The electronic computer represents one of our most spectacular examples of ephemeralization to date.

The first electronic computer ENIAC was built to calculate artillery-firing tables for the Army's Ballistic Research Laboratory in 1946. It contained 17,468 vacuum tubes, 7,200 crystal diodes, 1,500 relays, 10,000 capacitors and around 5 million hand-soldered joints. It weighed 27 tons, was roughly 8.5 by 3 by 80 feet, took up 680 square feet and consumed 150 KW of power. In its day it was a tour-de-force of computational power. Today, the processing chip in my digital wristwatch has orders of magnitude more computational power than ENIAC had, has more components and is about a quarter of the size of a postage stamp. And, comparing the central processing unit or CPU in the computer I'm using to write these words to ENIAC would be tantamount to comparing a sling shot to a hydrogen bomb. These developments are driving the exponential expansion and evolution of observational consciousness.

You have access to observational consciousness but you must share that access with all other observers and what differentiates observational consciousness from actual perception in any or all of the five perceptual realms is; it is the same for everyone.

Application of the scientific method creates observational consciousness in a highly structured and systematic manner by insisting that experimental results can be verified by any observer at any time.

Doing also depends on potentials. If you're going to do something you need choices, any one of which *could* be. I could get in my car and drive to the gym or to the library. Both possibilities exist in a kind of cloud of potential until one or the other is done. Staying with this example we come to understand doing as rotating space–time-mass into the universe to the gym or to the library. And, again, it is here, that the yin/yang or tensional aspect of doing is revealed; the mechanism by which the wholeness of being is torn. If I choose to drive to the gym I have the sense of missing out on what's going on at the library. This is a subtle but very real aspect of how potentials operate in terms of what we choose to do or not do. Potentials create concern and worry about what could have been and, in so doing, shatter the complete union of self and what is.

VISION AND TOUCH

Light can be thought of as either particles or waves traveling at c or 186,000 miles per second. Light, in addition to its velocity has another property called wavelength. For visible light, as opposed to x-rays or microwaves which are invisible, wavelength determines color. Because it's moving, one other property associated with light is momentum. In the visual realm of consciousness we experience a photon as a tiny point of light. In the tactile realm of consciousness we (our retinal cells) experience a photon as a minute touch. In this sense visual experience is reducible to tactile experience; the photon touches a retinal cell and we see a tiny point of light. Vision is reducible to touch. As a photon we *feel* light; as a wave we *see* it.

But, being reducible to touch isn't the same thing as being nothing more than touch or even being the same as touch. The reason is; saying light is nothing more than touch leaves out the particular order of light's constituent parts in the same sense that saying the Microsoft operating system, Windows7, is just a lot of ones and zeros on an optical disk. What makes light and the Windows7 operating system significant is the particular *order* of their constituent parts. Having said that, however, it is important to keep in mind that understanding the mechanism behind tactile reduction is what helps explain the structural relation between the visual and tactile realms of consciousness and their corresponding dimensions. Tactile reduction, in addition to explaining how touch becomes vision also explains the dual nature of light, the graininess of space–time, and mass energy equivalence.

The reason this is important is that it clarifies how and why visible light is the photon screen and how and why the photon screen is visual consciousness. Light is composed of quanta or photons arranged in dimensional patterns. Light is that facet of the quantum screen that becomes actualized in the form of vision. Every visual object is a pattern of photons.

The quantum screen represents the entirety of perceptual consciousness. Everything we hear smell, touch and taste is on the quantum screen. And, while the photon screen is a facet or part of the quantum screen, the quantum screen extends beyond the photon screen omni directionally to include hearing, smelling, taste and touch. Think of the quantum screen as a space–time extrapolation or projection of the photon screen. It is composed of a virtually infinite number of quanta—photon-sized bits of energy.

The image screen is not part of the photon screen. Observers appear on the quantum screen as objects but not just ordinary inanimate objects like bricks or baseballs. Observers are special case objects in that they are capable of creating order. But the order they create is not directly perceived in the same sense that we perceive a brick or a baseball. The order observers create is perceived by the *way* they move—the *way* they are able to keep from bumping into objects and other observers who happen to be in their way.

> A physical object is a pattern of quanta. The quantum screen is the physical world perceived directly.[17]

That's why it represents the entirety of perceptual consciousness.

THE IMAGE SCREEN

While perceptual consciousness is experienced on the quantum screen, observational consciousness is experienced on the image screen. But, unlike the quantum screen and its visual facet, the photon screen, the image screen is the context or framework of observational consciousness. The difference is important. Where the quantum screen is the physical world perceived directly, the image screen is what we experience *through* other observers. If you receive a letter from a friend, telling you about his trip to Tahiti, as you look at the paper it's written on and the ink used to write its contents and as you feel the weight and texture of the paper, all that is on the quantum screen. The physical properties of the letter are part of perceptual consciousness. The symbols your friend used to write the letter, the letters, words, punctuation marks and numbers, while they have physical properties, they also have order. It is that order which enables you to extract the meaning from the symbols and, in so doing, come to understand what your friend observed when he was in Tahiti. He may be describing what it was like to catch a large sailfish or how warm and clear the water was while he was snorkeling. And that's what's on the image screen. His experience becomes a *potential* experience for you. If you decided to go to Tahiti and visit the same places he did, you would be able to directly experience what he did. You could snorkel in the warm clear water and feel it on your skin, just as he did. You could go fishing and experience the exhilaration of catching your own sailfish. But, while reading the letter, all that is *potential* only because it's on the image screen and not on the quantum screen. In order to experience it on the quantum screen you must go to Tahiti.

The image screen is substantially different from the quantum screen in that it is *not* direct perceptual experience. Instead of being direct perceptual experience it is what is experienced *through* other observers. When you watch a baseball game or the Olympics on television you experience what other observers' experience. In this example, what you see and hear is on the quantum screen but, what's *happening* is on the image screen. If you listen to a baseball game on the radio, the sound coming from its speaker is on the quantum screen but the *game*—what's *happening* is on the image screen.

When you read a book, the paper and ink are on the quantum screen but the information contained in that book, whether it's a novel or a book about relativity, is accessed via the image screen.

If your friend comes to your house and while there, sitting in your living room, he tells you about his trip to Portugal, while you see him on the photon screen and hear him on the quantum screen, you access what he saw and experienced while he was in Portugal via the image screen.

I've provided these examples to demonstrate how and why the image screen *is* the world created by watching television, reading books, talking to other people directly, on the telephone or via Twitter and listening to what others have to say about what's going on in the world the galaxy or the cosmos. This is what *makes* the image screen the context or framework of observational consciousness.

The image screen doesn't have any physical properties in the same sense that order doesn't have any physical properties. Order is a context that you may or may not find related to physical objects. The image screen is the context or framework of observational consciousness.

The image screen is what anyone can see at any time or, put differently, it is potential perceptual consciousness as a whole.

> The quantum, photon and image screens are structurally identical because they are based on or dimensionally coordinated with the equality of one second of time and 186,000 miles of space or c—the speed of light. What differentiates the image screen from the quantum and photon screens is that it lacks the clarity and definition inherent in the quantum and photon screens.
>
> Where the photon screen is vision and the quantum screen perception as a whole, the image screen is the world beyond [or outside of] direct experience.[18]

In addition to being the context or framework of observational consciousness, the image screen is also the context or framework of "doable thought." Here's what I mean. If I want to fill my cup with coffee, I must see myself doing that or envision doing that on the image screen before I can accomplish the task. The same applies to driving to the library or the gym, going to bed or deciding to work or watch television. The moment I do any of these things they become perceptual consciousness. Anyone who happens to be where I am would be able to see what I've done. But, until I actually *do* it, filling my coffee cup, driving to the library or gym, going to bed, working or watching television remain on the image screen only and remain only potentials.

Doable thought and observational consciousness *share* the same context or framework. This suggests that, while the observational realm is expanding rapidly it is not yet fully developed or certainly not as fully developed as the visual, aural, tactile and chemical and olfactory realms or domains of consciousness are.

The dimensional structure of consciousness, in its entirety, is composed of the photon, quantum and image screens. What you see on the photon screen you can touch, hear, taste and smell on the quantum screen. On the image screen you can read about it, hear about it, or watch a presentation about it on television or on You Tube.

Reading about something you see on the photon screen is an example of where the evanescence or ephemeral nature of the screen is most apparent. Watching a video of something you see on the photon screen, like say a video about a new exercise machine you just bought and assembled which shows you how to use it, happens simultaneously on the photon, quantum and image screens. But the visual aspect of the video presentation *appears* as well defined *as if* you were watching someone demonstrate it using the physical machine you just assembled. Accessing a video presentation like this, instead of simply reading the manual makes the information on the image screen, in this instance, seem less ephemeral. What's important to keep in mind as you think about this example is that while watching the video presentation, even though it *appears* and in fact *is* less ephemeral than reading about it, what you're after and experiencing in both instances, is the information required to use the device. That information is what's on the image screen whether you access it by reading or watching a video even though the two methods of accessing that information are—subjectively—significantly different.

Potential experience is what links the photon, quantum and image screens but, in terms of how they are related to the dimensional structure of consciousness they serve as separate venues to a single entity—consciousness. Collectively they are the "screen."

Understanding the observational realm or domain of consciousness

The definition of the observational realm of consciousness is "potential perception." Put differently the observational realm of consciousness is the physical universe in its entirety because it includes everything that can be observed by anyone at anytime.

Everything I see you *could* see if you were standing where I am and vice versa. One clear implication of all this is the observational realm does not and cannot exist in isolation or absent observers. The engine that drives the constant evolution of the observational realm is the progress of science. To get at what's available, we access symbols on the quantum screen which appear as dimensional images, in the form of information, on the image screen.

Here, if we think carefully about how and why the dimensional structure of consciousness underlies how we experience the universe, we notice that the observational realm or domain is structured from and comes into being through perceptual experience in the same way that the visual realm of consciousness is structured from and comes into being through the tactile realm of consciousness.

In the same sense that vision is reducible to touch, communication is reducible to perception. If we expand our perspective to include everyone and everything in the universe, observational consciousness becomes a multi-observer experience or phenomenon made up of the perceptual experience of individual people. Viewed from that perspective, in a structural sense, it is the same thing as vision/light becoming a multicellular (whole organism) experience or phenomenon made up of the tactile experience of individual retinal cells.

The operational mechanics of the different modes through which we become aware of being embedded *in* consciousness instead of *having* consciousness loop back on and reinforce each other creating a wholeness of experience based on or modeled after light and the visual realm or domain of consciousness. That, too, is the reason space–time is in light rather than the other way around. Viewed from this perspective it becomes difficult, if not impossible to deny that the universe does, indeed, unfold in the touch of light.

No box...screen

As we began this examination of the dimensional structure of consciousness, I advised you that it would be painful; not because it is complicated or difficult but rather because it is so drastically different from the way you're accustomed to thinking about how the universe works and why we experience it in the ways that we do.

We've covered a lot of territory, most of it firmly attached to the physics that underlies the phenomena we've examined. And, in doing that, the screen becomes a much better explanatory model for the dimensional structure of consciousness than the box. I believe that while initially the screen model *appears* more complicated and tortuous than the box model, ultimately, it is simpler and has significantly more explanatory power.

Both models rest firmly on well-established physical principles. We didn't ignore or throw away the physics. And, even though that's true, it's important to keep in mind that there are some very real phenomena like dreams, thought, imagination, love, hate, envy, desire, *shadenfreude* and *ennui*, that are not on the screen.

Our examination of how the screen works and what it is—is firmly rooted in and restricted to science. Science does not concern itself with phenomena that do not appear on the screen because they are not objective. Using science to explore love or hate, envy, desire, *shadenfreude* or *ennui* would be as futile as trying to discover what's causing your diarrhea by examining the toilet in a very careful and disciplined manner.

Science restricts itself to the screen but, by definition, it cannot claim to cover all of reality even though, in the past, it has claimed to be able to do so.

Because of this the dimensional structure of consciousness is not a *scientific* theory. Its scope is enormously larger than science making it impossible to be tested within the bounds normally associated with science. Science not only restricts itself to the screen but also resolutely refuses to deal with phenomena that do not appear on the screen. For all these reasons, the dimensional structure of consciousness—the screen—even though it is firmly rooted in science—like the box—is a myth.

The key difference between the box and screen myths is the box myth cleaves firmly to the Cartesian belief in the existence of matter with or without observers. The cherry exists whether or not anyone ever sees tastes or smells it. The screen myth abandons that idea. Absent the seeing tasting or smelling—there is no cherry. Or, from the reverse perspective the cherry *is* the seeing tasting and smelling. For perceptual consciousness *what is*—is what you experience.

Look not at what you think or believe—look at what is!

Revisiting the abyss of solipsism

The clear implication of the dimensional structure of consciousness or the screen model of perceptual consciousness consisting of the quantum and photon screens is that what I see, smell, hear, taste or touch *depends* on my seeing, smelling, hearing, tasting or touching whatever *it* may be. Extending that logic, absent my seeing, smelling, hearing, tasting or touching any object; that object does not exist. In this sense and from this perspective, for every individual, consciousness is the ultimate reality and, since it is mine, everything is me. The entire universe is me.

Based on what we know and what science tells us the logic of this is practically indisputable. The abyss of solipsism looms large and, as we stare into the abyss, our moral, non-scientific and non-logical sensibilities are offended if, for no other reason, most of us have been taught, from an early age, that "the universe does not revolve around you." Unfortunately, like it or not, solipsism is and remains logical. The closer we choose to examine the idea of a world that exists independently of anyone observing it, the less we see. To be sure this is disquieting, in many ways repugnant, and repulsive to our moral sensibilities. At the same time, it becomes frustratingly complicated because, as I carefully examine my experience in and of the world, I discover that it's impossible for me to explain how the consciousness that seems to exist in others is, in some way, an aspect of my own consciousness. Somehow, it just doesn't scan. If you are part of me, why is it so hard for me to explain why you do what you choose to do? Shouldn't I know in advance and even be able to control what you, or for that matter, what anyone else does? Logically I should yet I discover, repeatedly, that I cannot. From this perspective solipsism becomes unacceptably complex.

As I examine this situation carefully, the problem seems to be consciousness; if I assume I have it and everyone else has it too. But, if I choose to back away from that assumption, then the problem is no longer consciousness. The problem, instead, becomes me. A more focused question then becomes; why does consciousness have to be me? If it's inside me or my brain it must be me but, if I and my brain are inside it, it does not have to be me. From this revised perspective, the assumed identity between self and consciousness vanishes.

One problem inherent in any discussion of consciousness is the word *consciousness*. Most people assume they know what consciousness means but its meaning is entangled with the connotation and etymology of other words like sentience and awareness. Here are some examples of what I mean:

When I Googled the word "sentient" I found:

- endowed with feeling and unstructured consciousness; "the living knew themselves just sentient puppets on God's stage."—T.E. Lawrence
- consciously perceiving; "sentient of the intolerable load," "a boy so sentient of his surroundings;" — W.A. White. wordnetweb.princeton.edu/pert/webwn

Sentience is the ability to feel or perceive subjectively. The term is used in philosophy (particularly in the philosophy of animal ethics and in...
en.wikipedia.org/wiki/Sentient

- Life form with the capability to feel sensation, such as pain; Conscious or aware; Experiencing sensation or feeling. (en.wikipedia.org/wiki/sentient)
- sentience—it may be conscious in the generic sense of simply being a sentient creature, one capable of sensing and responding to its world (Armstrong 1981). (www.illc.uva.nl/-seop/entries/consciousness/)

When I Googled the word "aware" I found:

- aware (p): (sometimes followed by 'of') having or showing knowledge or understanding or realization or perception; "was aware of his opponent's ...
- mindful: bearing in mind; attentive to; "ever mindful of her health"; "mindful of his responsibilities"; "mindful of these criticisms, I shall attempt to justify my action" (wordnetweb.princeton.edu/perl/webwn)

Awareness is the state or ability to perceive, to feel, or to be conscious of events, objects or sensory patterns. In this level of consciousness, sense data can be confirmed by an observer without necessarily understanding. ...
en.wikipedia.org/wiki/Aware

Vigilant or on one's guard against danger or difficulty; Conscious or having knowledge of something
en.wikipedia.org/wiki/aware

It's easy to see not only how but also how extreme the entanglement is between the words *consciousness*, *sentient*, and *aware* when you examine these definitions and notice how easily and often they substitute the word *consciousness* for the words *sentient* and *aware*.

There's a palpable slipperiness between the meanings of the words, how they are used and the context(s) in which they are used.

Among other things, this indicates the extent to which we understand, as well as the extent to which we do not understand what they mean.

They are, after all, bona fide definitions. But they morph and interpenetrate in ways that are both interesting and confusing.

If nothing else, I believe it's important to become and remain mindful of this as you continue to explore and think about the dimensional structure of consciousness, what it means and how it works.

Space and time are structures of consciousness. "Matter" or what we assume to be "matter," is an actualization within that structure. If this is true, as I believe it is, consciousness is not within my brain or me and there is clearly no identity between self and consciousness.

The perspective of *perceptual* consciousness is self. The reason we identify self with consciousness is everything we see, hear, smell, taste, or touch, is from the perspective of self. This is the constantly operational, *always on*, mechanism that closes the understanding loop and makes it feel *as if* everyone *has* consciousness.

The new model—the dimensional structure of consciousness—yields a new perspective. Instead of everyone *having* consciousness; everyone is *in* consciousness. This perspective severs the identity between self and consciousness.

Everything I've said, so far, applies to *perceptual* consciousness. Everything we perceive using our senses is, by definition, from the perspective of self. It happens in our heads. However, the same self that *is* the perspective of *perceptual* consciousness *is not* the perspective of *observational* consciousness because, by definition, observational consciousness is a totality over and above the separate individuals or *selves* that create and maintain it. The order associated with and embedded in observational consciousness goes beyond its essentials and, in that sense, becomes a reality more fundamental than what it contains.

Every individual experiences observational consciousness from the perspective of self because it is reduced to sights and sounds and other sensations that are *perceived* by self. From the standpoint or perspective of self, the only way observational consciousness can possibly be experienced is, by definition, in its *reduced* form. And while that's true because of the way the dimensional structure of consciousness operates, observational consciousness becomes and remains independent of any individual self whether that self is yours, mine or anybody else's.

If we accept this definition and add an appreciation of what observational consciousness is, how it is structured and how it works we create the epistemological and moral safety required to accept perceptual consciousness, in its entirety, as solipsism. If we come to understand that perceptual consciousness is not the whole of consciousness we may comfortably allow for the existence of only one perceptual consciousness. We may safely jump into the abyss of solipsism if we understand we are not doing so, empty handed.

Observational consciousness becomes the backstop which provides a bottom to the previously bottomless abyss of solipsism and the reason is that while, for me, it's true everything I see, hear, smell taste and touch is perceptual it *is not* true that everything he, she or they see, hear, smell taste and touch is perceptual because solipsism disallows that.

One way others have tried to explain, why other people or observers appear to be able to do the same things I do, like seeing hearing, tasting etc. within the framework of solipsism is to suggest that the other observers I see are nothing more than extremely complex feedback mechanisms capable of reacting to various kinds of stimuli and capable of informing me about objects that I cannot observe directly like when they tell me there's a gun on the coffee table in the living room when I'm not in the living room. While it is true there is a certain logic to this explanation it is also true that it is undeniably absurd, immoral and hideously complicated. For all these reasons, most people, when presented with this explanation simply reject it out of hand. The explanation is so ugly and bizarre it's impossible to take it seriously.

There are five domains of perceptual consciousness but there are also additional domains or realms of consciousness which are *not* perceptual. In thinking about how this works it's helpful to keep in mind how the word *self* acts as a kind of pivot point between the perceptual and observational realms and tends to morph in meaning as you switch from realm to realm by forcing you to reexamine and change your perspective of what it is that you're trying to "see."

In working through this mental gymnastic, himself, herself or yourself becomes and remains a perspective or standpoint of observational information. Here's what I mean by that. If he or she says they see a man with a gun, at a particular point in space-time; if I were standing where he or she is standing, I too would see the man with the gun. What he or she says they see, in this sense becomes a locus of order. It does not, however, become a separate perceptual consciousness. And, while it's true that he or she do things that both create order and help to keep them alive, it's also true that they only *appear* to have consciousness because, like me, instead of *having* consciousness, they are *in* consciousness.

What this implies is the difference between *my* self and all other selves, is distinctly physical as well as structural. Every self is the pivot or focal point of *doing* but only *my* self includes perceptual consciousness. The clear implication of this is that *all* selves, in the observational realm or domain of consciousness, including *my* self, are equivalent.

When I tell you what I see and feel, and when I listen to what it is you say that you see and feel, *your* self and *my* self become equivalent both morally and physically. From this perspective it is only the moral and physical equivalence of individual selves that allows for and brings observational consciousness into existence.

It is in this sense that observational consciousness becomes *potential* perceptual consciousness. If I say I see a man with a gun, then you or anyone else, standing where I am standing must also see a man with a gun.

For me, in the perceptual realms, "you" become *just* an orderly arrangement of symbols representing potential perception. But, when we add being into the mix of what we are considering and examining we discover it is not limited to perceptual consciousness. When we accede to the idea that all selves are morally equivalent; being becomes disentangled from what only "I" see or feel and instead becomes firmly embedded in what I *potentially* see or feel, based on what you tell me, whether you do that face-to-face, in a letter, in an email, or over the telephone. And this is the mechanism through which potential perception becomes as real as actual perception; the mechanism, through which, observational consciousness is created and sustained.

Observational consciousness vanquishes the hideousness associated with solipsism without banishing it or denying it exists and provides immediate and continuous access to empathy. It allows me to feel your pain without perceiving it directly. It allows your pain to become as *real* for me as it is for you even though I do not experience it in any perceptual realm.

When you share a perceptual experience with me, while the *perceptual* experience is not actualized for me what you tell me about that experience is. In this sense, observational consciousness, while not allowing me to "become" you does allow "us" to become a being greater than both of us, separately.

This understanding of what observational consciousness is, how it comes into being, how it works, why it vanquishes the ugliness associated with solipsism, and how it enables empathy is a secular appreciation of what religion teaches us about your experience being as real as mine is. It points, unambiguously, to a great truth which is impossible to perceive directly. The only way we can know about it is through being.

The tension that exists between self and others—between actual and potential perception is not between one consciousness and another consciousness—between your consciousness and my consciousness. It is a tension between one level and another of the *same* consciousness we all find ourselves *in*. It codifies the advice found in most of the great religions—do unto others as you would have them do unto you.

The tension that exists between perceptual and observational consciousness; and thus, between self and others is the foundation for understanding what we know of the world. We have, so far, relied on perceptual consciousness to help us understand and make sense of the world, what it is and how it works. I'll believe it when I see it became a kind of standard we used to verify both reality and reliability as we examined our surroundings and tried to make sense of our experiences. But that's beginning to change as we come to rely more on potential rather than actual perception to make sense of the world.

The realm of observational consciousness is expanding at an exponential rate facilitated by the internet, cell phones and web based applications like Google, Facebook, Myspace, and Twitter. Today, we are able to learn about events happening around the world and, even in outer space, almost instantaneously. When an earthquake, tsunami or hurricane happens anywhere on the globe; people not only find out about it immediately, they're also able to determine, just as quickly, whether loved ones in the vicinity of the disaster, were involved or impacted by what happened. This was not so in the twentieth century because while we were able to communicate over vast distances, our ability to do that was much more limited than it is now with the vast access almost everybody has via the world wide web using applications like Twitter, Facebook and Myspace.

Using cell phones we are able to send pictures and video instantaneously of events, as they occur. This capability alone has had and is continuing to have an enormous impact on how we live our lives and, more importantly, how we relate to each other.

Ten years ago, unless television news cameras recorded the incident, as they did during the Rodney King beating in Los Angeles, no one would have known about how the police conducted themselves and behaved. It might have been reported on, after the incident, in the newspapers but, even then, the overall and immediate impact would have been severely attenuated. The only reason the incident had the enormous impact it did is because television news cameras happened to be there when it happened. Today, citizens with cell phones regularly record misbehavior by police and military personnel and share the video of what happened on the web through YouTube and other web based video sharing web sites. This rapid expansion in observational consciousness is exerting strong and relentless pressure on government police and military personnel across the globe in ways that wouldn't have even been possible a decade ago.

The political landscape is changing globally forcing regimes which, in the past, easily controlled the information available to their citizens, to rethink their options. And, while it's true, all this is being facilitated by vastly accelerated and improved communication capabilities, what's really driving the rapid global changes we are all experiencing is the simultaneous expansion of observational consciousness on a scale never before experienced in human history.

As observational consciousness expands—power shifts.

The ascendance of potential perception

As observational consciousness expands, we come to rely increasingly on potential rather than actual perception. If we see a hornet's nest we avoid touching it, not because the hornets are stinging us but, because we do not want to be stung.

If, before crossing the street, we see a Corvette speeding toward us at 90 miles per hour, we wait until it has passed before stepping into the intersection because we know it is really there and it would hurt us if we stepped into its path. We avoid poison ivy not because we feel the itching it would cause if we touched it but because we want to avoid experiencing that unpleasant tactile sensation.

These are concrete examples of the ways in which we limit and control our *actual* experiences through a series of ongoing complex dimensional interchanges that allow us to actualize desirable experiences in the perceptual realms and avoid experiences which *would be* undesirable or fatal. This is how we stay alive. We have learned that potential perception is as real as actual perception, even in the perceptual realms.

This doesn't change what the perceptual realms are or how they work. The lamp you see sitting on your desk is really there because you will really touch it if you reach your hand out to the space and time it occupies. But as we learn more about observational consciousness and continue to evolve in terms of our understanding of how the dimensional structure of consciousness works as it relates to perceptual consciousness, potential perceptual consciousness becomes increasingly more important than actual perceptual consciousness.

We may not think much about it but, in truth, the importance of *potential* perception is central to who we are and what we do. Most of us get up every morning, get dressed, and go to work. If you put on work clothes and get in your truck to drive to the construction site; if you're a carpenter, when you get there, you get your tools and begin attending to tasks like framing the houses you're working on in the new housing project. You might use a nail gun and a chop saw. As you think about all these things you do, it's interesting to note there's no money in any of them. While you're working you're not experiencing the things you really want to experience. You use the nail gun and chop saw in order to get paid. And, even after you've been paid, the money you get is not actual food or shelter or entertainment. If you want any of those things you have to buy them. If you want to watch premium channels on your television set, you have to pay the cable bill before you can watch them. What you're doing, in all this, is reacting to potential rather than actual experience.

While it is true that we do not experience the *actual* perceptions of other living beings or observers, that doesn't mean they are not alive. When we accept the existence and reality of solipsism as it relates to perceptual consciousness that does not mean other living beings or observers are not alive or that they have no feelings. The dimensional structure of consciousness—the screen—works the same way for them as it does for me and, for that reason, every observer, whether it's another person, a dog, an amoeba, a tree or a blade of grass and even every cell in that blade of grass, is an entire world of sensation and feeling. Every friend, family member, foe, lover and significant other is a universe of stars, galaxies, comets, auroras, and shooting stars equivalent to what you see with your own eyes. Their experience is not actualized for you only in the sense that it is not available to you in the perceptual realms because *only* your experience is available to you in the perceptual realms. You learn of their experiences through the facility of language. This allows you to see what they are seeing without having to go where they are. Language allows their experience to become yours without that experience becoming *actualized* for you in the perceptual realms. Perceptual solipsism exists but it doesn't change the fact that:

> "When we live with and love other people, potential perception is as real as direct perception."[19]

And here, with this understanding, we find—once again—a secular appreciation of truth which enables empathy and codifies the most fundamental tenet of all religions—do unto others as you would have them do unto you. The admonition becomes and remains supremely logical with or without religion and regardless of whether or not you choose to invoke a belief in God.

Making a hole in the photon screen

In the same sense that the quantum screen is an extension or extrapolation of the photon screen, the image screen is an extrapolation of both the photon and quantum screens and is based on the same dimensional structure as the photon and quantum screens are. If we examine them separately we find:

The photon screen is four-dimensional based on space, which has three space dimensions, length, width, height, and also includes time.

The quantum screen is five-dimensional based on length, width, height and time but also includes mass which, as identified earlier is space–time-time.

The image screen is six-dimensional based on length, width, height, time and mass (space–time-time) but also includes order.

As we experience all three screens, if we think about what they are, how they work and how we actually experience them we notice that they all coincide. But we also notice something else about how we experience what happens on each screen.

The photon screen is immediate and finely grained because photons move very fast, have no mass and constitute the field of vision or visual consciousness. It enables us to *see* the world.

The quantum screen is actual perception, in its entirety and, in addition to vision includes, hearing, smell, taste and touch which are not as immediate as vision. Everything we experience on the photon and quantum screens becomes actualized for us in the form of perceptions in one or more of the perceptual realms.

The image screen, while based on the same dimensional structure as the photon and quantum screens is different in terms of how we experience what is on it. It is substantially more abstract than the photon or quantum screens; not only because, it lacks their immediacy but also because what we experience on the image screen requires an extra step before we can gain access to it—we must *envision* what is on the image screen in order to gain access to what is on it.

When my friend calls me from Tahiti to regale me with the experience he had catching a two hundred pound sailfish, I must *envision* him doing that in order to extract the information he is sharing with me. The image screen is what's down the street, in the basement, in the back yard or on the moon and, in that sense, is the *context* of, what for me, becomes *potential* rather than *actual* perception.

To get a feeling for how the photon, quantum and image screens work and how we access them imagine that my friend, who called me from Tahiti last week, is now in my living room with me. He's sitting on the couch and I'm in my easy chair. As I look at him I see him, the couch, the coffee table in front of the couch and the lamp, to his left on the photon screen which is only a part of the quantum screen. As I listen again to his story about catching the sailfish, while I watch him on the photon screen, I hear what he's telling me and smell the smoke from his cigarette on the quantum screen.

If, while I'm listening to his story about the sailfish, I hear a chittering sound coming from the back yard that too is on the quantum screen along with the taste of the fine single malt scotch I sipped while listening to what he was telling me.

I do not see the squirrel making the chittering sound in my back yard, but I hear it in space and time. If I turn my head to look out the window at the squirrel; the photons that make up my friend, on the couch, are no longer actualized for me in the visual realm. Instead, I see the squirrel in my back yard and while I no longer "see" my friend he's still on the quantum screen because I'm still listening to him and I can still smell his cigarette smoke, so I know he hasn't left the room. In this sense he remains a quantum pattern. If I turn my head back in the direction of my friend, he will re actualize, for me, as photons in the visual realm or domain. As we examine the photon screen versus the quantum screen, what becomes most immediately apparent is the photon screen is spectacularly vivid and detailed. And, while the quantum screen lacks the vibrancy and detail inherent in the photon screen, we notice it is much larger and more inclusive.

When I stopped looking at the squirrel and turned my eyes back toward my friend, the image I saw of him on the couch was much clearer than any image of him I could only hear or smell.

If we stay with our exploration of the screens and how they work in this little scenario about my friend and me, it becomes obvious that it is impossible for me to look at the squirrel and my friend simultaneously. I could certainly look at him, listen to what he's telling me and sip my single malt scotch simultaneously but, while doing all that, the only experience available to me, of the squirrel, is the chittering sound I heard outside in my backyard. It's not possible for me to look at my friend and the squirrel at the same time. And, if while listening to the squirrel chittering I also hear a helicopter, in the sky and my neighbor's dog comes into my backyard to bark at the squirrel, I'm unable to look at any of those things while looking at my friend. I know they exist in space and time because I'm experiencing them on the quantum screen but, if I want to see them, I must scroll the photon screen away from my friend, through the window and into the backyard allowing me to add visual (photon screen) information to what I hear on the quantum screen. When I do that I get much more vivid information about what is going on out there.

My friend's description of how he caught the sailfish is fascinating. As I listen to him, I can almost feel what it would be like to hook into a two hundred pound sailfish and hear the gears, in my fishing pole reel scream as the fish runs with the bait in his mouth while I, strapped into my chair pull hard against the line, the muscles in my arms and back burning with the effort. But all that is on the image screen, in the observational realm, and happening in my mind.

If I really wanted to, I could move the quantum screen to Tahiti and charter a fishing boat. If I did, I could see, hear, feel and smell what my friend did in the perceptual realms. The experience would be much more vibrant than the one I'm having on the image screen. But I can't go everywhere and do everything I hear or read about because while I'd like to there are too many places to go and things to do and not enough time.

Compared to the quantum screen, the image screen is substantially larger. Because that's so, it is and always will be impossible for me to actualize everything on the image screen. So, whether I want to or not, I'm forced to be choosy about where I decide to scroll the quantum screen onto the image screen. I take in direct perceptual experience when and where I can as time and circumstances allow. For everything else, I'm forced to settle for observational consciousness.

This story about my friend and I demonstrates key aspects of the dimensional structure of consciousness, what the photon, quantum and image screens are, how they work and how we experience them. It also demonstrates how and why, for most of us, the observational realm via the image screen is the venue through which we experience *most* of what we know about what's going on in the world. It also reveals that the quantum screen is a perceptually actualized extrapolation or facet of the image screen in the same sense that the photon screen is a perceptually actualized extrapolation or facet of the quantum screen.

All three screens are fundamentally based on the dimensional structure of consciousness but each screen, starting with the photon screen and finishing with the image screen adds an additional dimension with the photon screen having four dimensions, the quantum screen having five dimensions and the image screen having six.

What we experience through the image screen is not as clear or well defined as what we experience through the quantum and photon screens because the image screen is not as fully developed as the quantum and photon screens and because, in order to *get at* what's on the image screen we must perform an additional step of *envisioning* the information available on the image screen.

The observational realm of consciousness is expanding in importance at an exponential rate, as I mentioned earlier. One indication of this spike in importance is the extent to which we've become attached to our cell phones and, also, the extent to which we are opting for *smart* phones versus traditional cell phones which only allowed us to make phone calls. Another indication is a dangerous new societal phenomenon, texting while driving which threatens the life of the person texting as well as the lives of others, on the road, in the vicinity of anyone who texts while driving.

As important as the observational realm is becoming, what we experience on the image screen, even while using our smart phones, is not as vivid or well defined as what we experience in the *direct* perceptual realms of consciousness.

Imagine that you and I are in my house. You are in the kitchen and I am in the living room. I ask you to bring me the set of small jeweler's screwdrivers so I can tighten the screws on my glasses. When you ask me where they are I say in the kitchen cabinet. Since there are quite a few kitchen cabinets you ask which cabinet? I tell you the one above and to the right of the stove.

You open the cabinet and tell me you don't see any screwdrivers. You experience an observational image of the screwdrivers on your image screen in the cabinet I told you they were in but, when you look inside, you don't see them. I ask you to rummage around and, even when you do, you still can't find the screwdrivers. After a while, I lose patience come into the kitchen and realize you're looking in the wrong cabinet. I open the one to the right of the one you were looking in and point to the screwdrivers.

You say, "You said the cabinet above and to the right of the stove." To which I reply, "it is above and to the right of the stove." If you were my good friend Tommy we'd exchange a number of nasty expletives and laugh but the point of all this is photons are a lot more precise than words.

One reason Samuel Avery chose the term image screen instead of observation screen to describe what we experience in the observational realm of consciousness is precisely because photons are more specific than words; especially in a scenario like the one regarding the jeweler's screwdrivers.

Another reason is we use the same screen for other kinds of nonperceptual experience. I experienced my friend's description of catching the sailfish in Tahiti the same way I would experience an elephant dancing on the head of a pin or what it would be like if I could walk through walls. And, because we use the same screen for different kinds of images, we're tempted to assume that when I envision my friend catching the sailfish, I'm imagining it. But that's a serious mistake. It's wrong because I didn't imagine it. Unlike an elephant dancing on the head of a pin or me walking through walls, my friend really did catch a sailfish in Tahiti. Tahiti really exists and, if I wanted to, I could go there and catch my own sailfish. What causes the confusion is I use the same screen to experience catching the sailfish as I do to imagine the elephant dancing on the head of a pin and what it would be like to walk through walls. This is a subtle but important aspect of, not only what the image screen is and how it works, but also of how we use it. We do this all the time but we don't think much about it.

There is yet another subtlety regarding the image screen and how we use it. In addition to using it to envision my friend catching a sailfish in Tahiti, which isn't imagination or an elephant dancing on the head of a pin which is imagination, I also use it to envision doable thought.

Here's what I mean. If my friend tells me he brought a bottle of our favorite single malt scotch and it's in his bag on the bed in the guest bedroom; and he asks me to bring it into the living room, I envision the bottle where he told me it was in the same context as I envision going into the guest bedroom to get the bottle. Neither envisioning the bottle nor envisioning going to get it are perceptual experiences but both are what *could be* perceived. What's important to think about and understand about this is envisioning where the bottle is and envisioning going to get it were both experienced on a screen (the image screen) which, while not in space–time is structured on and coordinated with space–time.

As you think about this, it becomes apparent that the observational realm of consciousness and doable thought share the same context. And, while this is true, it's important to understand that the context they share is *order*. But, of equal importance, observation is not what is doable because, what is doable usually begins as just one possibility among at least one additional, and in some cases, many more possibilities that exist as a sort of cloud of *potentials*.

When you think about what it is you're going to do, like getting up from your chair to get some potato chips or waiting until the next commercial comes on during the show you're watching on television, you're experiencing those thoughts in the dimension of order before you actually bring the action into the perceptual realms. Nobody, not even you, can see, or experience, in any of the perceptual realms, what you're experiencing in that dimension of order *until* or *unless* you do it.

Once you do it—get up and get the potato chips—you coordinate the order of that doable thought with space and time making it possible for everyone, including you, to experience what you *thought about doing*; in the perceptual realms. Others, in the room, will be able to see you get up from your chair, go into the kitchen and return with the potato chips. That is the definition of doing. What you thought about doing, once you do it, becomes what everybody can see, hear, smell, taste or touch.

As we explore the image screen, we find that in addition to observation, other kinds of non-perceptual images appear on it like doable thought and elephants dancing on the head of a pin. And while this makes the concept of the image screen, its existence, and how it works, confusing it also provides some insight into how the perceptual realms or domains of consciousness have evolved, over time.

Before the tactile and olfactory realms came into existence; they had to evolve from the most fundamental or primitive realm of consciousness, the chemical realm. And, before the auditory and visual realms came into existence they had to evolve from the more primitive tactile realm of consciousness. As this evolution took place periods of differentiation had to be endured. It's doubtful that our ability to smell and, hence, the olfactory realm of consciousness, suddenly popped into existence fully developed and differentiated from taste without an intervening period of time over which it could expand and find its own perceptual niche. And, the same is true for the visual and auditory realms of consciousness. A period of evolutionary time was required before the tactile realm of consciousness or touch could develop into our ability to hear and see. As each new realm developed, it had to be experienced within the old realm until, eventually; it revealed or discovered its own dimensionality.

This is what's happening now with the observational realm of consciousness. As it continues to endure its own period of differentiation, it will continue to separate further from perceptual consciousness by more firmly establishing order as its corresponding dimension.

We are witnessing a spike in the further development of the observational realm and, as we participate in bringing about its evolution the difference between the observational and perceptual realms is becoming more apparent because a separate context for the observational realm is rapidly evolving too.

In the past observation depended almost exclusively on language. That is no longer true. Now we have television, computers, YouTube, Google, Facebook, My Space, Twitter, and smart phones with high resolution video cameras built in. We can manipulate the information available to us, on our smart phones, by touching the electronic display in certain ways. Swiping motions flip virtual pages of information and taps on icons access different operational modes.

When we look at images on a television screen, or for that matter, any kind of electronic display, although we may not think about it, the images are not in real space, in the sense that the coffee cup on my desk is in real space. Instead, they are composed of a simple two-dimensional array of pixels or picture elements or dots that *look like* real space. This array or pattern of pixels is an orderly arrangement, on the two-dimensional space of the display that makes us think of or *envision* real three dimensional space. After all, an electronic display only has two dimensions.

When we watch Jack Bauer on 24, chase a bad guy into an abandoned warehouse, what we're really *seeing* is not Jack Bauer but millions of colored pixels. His gun is not a real perceptual object even though we do see it perceptually. But, just like Jack, his gun is just an orderly array of colored pixels on a two-dimensional display. And, even though it *looks like* a gun we *would* see in real space–time it's important to understand that it is not.

We watch Jack Bauer and any other television show in observational consciousness that *appears* to exist in a space and time context which *looks like* it is visual but, it is not visual. What we're really doing is *envisioning* Jack Bauer chasing the bad guy and shooting at him on the television screen—not in real space–time—the same way that anybody else can envision him there.

The key difference here is what we see on a television screen, can be seen by anyone and is, therefore, by definition, observational as opposed to being perceptual. This is very different from you or me sitting in our living room looking at a coffee cup sitting on the table. My experience of the coffee cup is mine alone as is yours.

But, with Jack Bauer, while we gain access to him perceptually, through the visual and aural realms of consciousness, what we're gaining access to is Jack and Jack, as we watch him on the screen, unlike the coffee cup, exists in *observational* consciousness.

Another way to think about this is, if we were reading the story about Jack chasing the bad guy into the abandoned warehouse, our experience of him would be much less vivid than when we watch him on the television screen. But, in both instances, Jack exists in observational consciousness. And, in both instances, because Jack is an *observational* image he can be and is accessed simultaneously by hundreds, thousands or millions of people. As they watch him, on television, they all assume they're seeing him, in real space and time but they're not.

What makes what's available or *seen* on a television screen *observational* and not *perceptual* is—by definition—it can be and is seen simultaneously by millions of people. That's very different from what happens when I look at the squirrel in my back yard. I see the squirrel on the photon screen, in real space and time. And, while it's certainly true that if you were in the room and came over to the window, you too would experience the squirrel on the photon screen in real space and time. But, you don't have to think about this too long or too hard to realize there is a very real physical limit on how many people could experience the squirrel in my back yard in real space and time—simultaneously.

If I had the equipment to broadcast the image of that squirrel to the television show Animal Planet, then millions of people would be able to see him but, if they did, they wouldn't be seeing him in real space and time—they'd be *envisioning* him, on the image screen while *seeing* tiny dots perform an extremely *orderly* dance on the television screen. Because this is so, a television screen is, in a very immediate and real sense, a souped-up version of the image screen.

Observational consciousness existed long before television was invented or became generally available to the population at large. But, as we adopted and adapted to television, it began to evolve into a separate context for the observational realm of consciousness, in its entirety.

When we watch the Olympic games or the Super Bowl on television, we share the commonplace experience of hundreds of millions of people. Every person watching is seeing the same images from exactly the same perspective.

Here's the part about the hole in the photon screen. If you think about it carefully, a television screen is really a *hole* in the photon screen that reveals the image screen in a format that *appears* perceptual but is not. When we watch the Olympics, the Super Bowl or any other television broadcast, what we *see* in the visual realm of perception is pixels doing an orderly dance on the screen which causes us to *envision* the action on the image screen. For the individual(s) watching, the screen exists in real space-time. What's happening, on the screen, does not because they're not where the action is occurring. The sound coming from the speakers, like the television screen, is also in real space-time but, once again, it functions in much the same way that my friend's words did, when he described what it was like to catch that sailfish in Tahiti. The sound becomes another perceptual portal to the image screen which helps to enhance the order we are envisioning on the image screen.

Everything I've said about the television screen applies equally to other kinds of electronic displays when they're used in the same way. Millions of people communicating with each other via Facebook, My Space or Twitter, whether the device they're using is a PC or a smart phone are gaining direct access to the image screen through a hole in the photon screen. Their display becomes the hole.

As television screens and other kinds of electronic imaging modalities advance, improve, and become even more highly dimensional they will ultimately evolve into "the observation screen." As that happens; the observational realm of consciousness will become more differentiated and highly defined than it is now.

An example of the ultimate manifestation of the observational realm becoming so highly evolved it would be difficult to distinguish it from perceptual consciousness in real space-time, is the holodeck on Star Trek. Once you enter the holodeck and invoke the program, with which you wish to interact, all the objects and people you see and interact with are indistinguishable from experiences you have in real space-time. We're not there yet but even with games we play using devices like the X Box or the Sony PS 3 or the WII gaming systems, the experiences become so immersive they take on a reality that, for some, becomes addictive.

This is the mechanism through which the observational realm of consciousness is evolving. By using these kinds of devices; adopting them and adapting to what they have to offer, we are participating directly in that evolution.

The dimensional structure of consciousness is a model or paradigm which provides significant explanatory power regarding the constantly operational—always on—mechanisms which allow us to experience the universe and what's going on around us. Because this paradigm is so well structured if you think about it carefully and consider the experiences you have in the perceptual and observational realms of consciousness, it becomes increasingly difficult to hold onto the *box* paradigm of how the universe works and what it truly is. It's as if the dimensional structure of consciousness is, itself, *becoming* a realm or domain of consciousness. And while it's certainly true part of the reason this is happening is because of the model's accuracy and explanatory power, another equally important reason is we are using the same dimensional structure we experience in the perceptual realms to create and, ultimately be able to become aware of and perceive the model and bring some understanding to how it works and why it works so well.

The screen model provides a complete and sufficient explanation for direct perception or what we experience in the perceptual realms of consciousness and why. For the observational realm of consciousness, the model or paradigm is evolving into the thing itself.

The flow of observational consciousness

Observational consciousness is, by definition, dynamic. It is something observers experience through doing—it flows between and amongst observers when and if the circumstances exist which allow or provide for that flow. One such circumstance is spoken language. Others include books and magazines, television, radio, telephones and cell phones, computers and applications like Google, Digg.com, Twitter, Facebook and My Space. As important as all these circumstances or venues which provide for and enhance our ability to communicate and access the observational realm may be, it is even more important to understand that none of them, alone or collectively, *are* observational consciousness. Observational consciousness *flows through* them. Language, books, television etc. are dimensional manifestations of observational consciousness.

If we tighten our focus here to include only the symbols which make language possible, whether letters and numbers or spoken sounds coming from other observers directly or through devices of some sort, we come to appreciate they exist in the five dimensions associated with perceptual consciousness and are also *arranged* in an additional dimension—*order*.

As you read these words, observational consciousness flows through them because and only if you have a working command of English. It's important to understand, however, that observational consciousness would not flow through these words if you were a cricket, a squirrel, a dog or if the only languages you could speak, read and write in, were Farsi and Urdu. And, while it's true, observational consciousness potentially exists in and for all observers it is not actualized in all observers unless they share certain capabilities like the ability to read, write and understand certain languages.

As things are now, observational consciousness is not universal because people, in the world, speak different languages. People who speak German represent a separate collective self from those who speak Italian, Polish, Urdu, Farsi, Hindi,

Chinese or Portuguese. Those who speak English are liberally sprinkled amongst those who speak other languages and English is spoken by many more people than by those who exclusively speak languages other than English. But, even so, English is not universal.

Until and unless a universal language ultimately emerges, collectively, we'll be confronted, in terms of observational consciousness, with what the prison boss in the movie Cool Hand Luke famously observed—*what we have here is failure to communicate.*

Until recently, language was the linchpin of observational consciousness. If a universal language existed, for humans, observational consciousness would be much closer to being universal than it is now. But, even with the existence of a universal language for all humans, other observers like, dogs, frogs, lizards, bears, gorillas, whales fish, penguins, insects, algae and bacteria would be left out. Observational consciousness will never be universal for *all* observers.

Eventually, observational consciousness will become universal for humans as we continue to move away from language as the primary facilitator or linchpin of observational consciousness. As powerful and important as language has been and still is, it is being rapidly supplanted by an array of electronic devices like radio, television, computers, cell phones, I Pods, I Pads and e-books which are more powerful than language alone. As we continue to adopt and adapt to these devices and as their availability and variety continues to expand, the same collective self will eventually become actualized in all human observers. You and I, our loved ones, friends and enemies will become entangled in a larger self in whom everybody else is entangled. And while some of these devices will still rely on language to facilitate communication and the flow of observational consciousness, others will rely instead on music, numbers, images and direct visual experiences which transcend the rhetorical, syntactic, informational, and national limitations inherent in language alone.

Ultimately, who and what we are is shaped, not only by what we see, but also by how we interpret and react to what we see. We become, not only what we behold, but to a large extent, how well we are able to interpret what we behold before deciding whether, how, or if we should react.

Language and symbols are important in helping us to understand what is happening around where we are and in places where we are not. When I watched images of hurricane Katrina devastating Louisiana, on television, so did you and millions of other people around the world. The same is true of the BP oil well that exploded in the Gulf of Mexico releasing millions of gallons of crude oil into the water. We didn't need symbols to tell us that Louisiana was being devastated or that an oil spill representing an epic environmental disaster was occurring. We could see it for ourselves. The ever-constant *potential* for information about what's happening, at any moment, anywhere in the world, is the television screen. To gain access to that information we need do nothing more than turn on the television and watch what's happening on the screen. With the availability of smart phones, many of us carry that television screen around in our pockets.

Language always was and always will be an important venue through which we gain access to observational consciousness and through which observational consciousness flows. But, as we continue to adopt and adapt to television and carry it with us wherever we go, the flow of observational consciousness will become increasingly diverted to the television screen.

Television is still a relatively new technology. As a society, we began to embrace it by bringing it into our living rooms in the late 1940s.

"In 1950 fewer than 10 percent of American households had a television set. A decade later fewer than 10 percent of American households were without one. In a single decade America changed from a land of newspapers and radios to a nation dominated by the TV screen. It wasn't just the number of televisions that increased, but the time Americans spent with the medium as well."[20]

That's a truly astonishing statistic. The scale of television's penetration into society was unprecedented. And thus we began to change our collective preference for the medium through which observational consciousness flows.

By the mid to late 1950s, a new societal problem began to surface encapsulated by the question "Why can't Johnny read?" We began to notice that children were losing their ability to learn how to read, and many theories surfaced regarding why that was so. Rudolph Flesch wrote a book that focused on how reading was being taught and suggested ways to improve the methods used to teach reading skills.

Many articles were written on the subject but one thing that seems clear, in retrospect, is that in the mid 1950s, by the time Johnny got into first grade, he had already logged thousands of hours in front of the television screen. While watching television he saw everything from drama and variety shows to children's programs and advertisements for everything from laundry detergent and toothpaste to cigarettes and shampoo. So, when the teacher presented him with his first primer with stories about Dick and Jane and their dog Spot with dialog like:

> See Dick.
> See Jane.
> See Spot.
> See Spot run.

Johnny's reaction was, where's the Lone Ranger? Where's Howdy Doody? And later, where's Kojack?

Johnny had become accustomed to the fluid immediacy associated with observational consciousness flowing through the television screen and the richness of information available via that medium. He had already learned more about the world and how it works than children ten years earlier could ever have possibly known. It became increasingly difficult and ultimately impossible for Johnny to identify with the adventures of Dick, Jane and Spot and, as a result, he rejected reading in favor of television.

Literacy statistics in the United States today are grim. According to the National Right to Read Foundation, 42 million Americans can't read at all; 50 million are unable to read at a higher level than is expected of a fourth or fifth grader.

The number of adults that are classified as functionally illiterate increases by about 2.25 million each year.

Of high school seniors 20 percent can be classified as being functionally illiterate at the time they graduate. This situation isn't getting any better. In the

1940s and through the late 1950s, students in primary and high school were expected to learn the fundamentals of English rhetoric, grammar, syntax, sentence structure, spelling and writing. By the mid 70s overall skill levels had begun to deteriorate precipitously.

I remember being astonished when, as the administrative director of a radiology department, in Chicago, I began to review applications for the position of director of the radiologic technology program associated with that department. All the applicants had baccalaureate degrees and yet the sentences they wrote in their applications were grammatically, syntactically and rhetorically deficient. I was seeing, first hand, the implications of the stories I'd read about "why Johnny can't read."

Faced with choosing television as a means of accessing observational consciousness, versus reading, we opt for the television screen. Not because we're lazy but rather because it's so ubiquitous, fast and fluid compared to reading. But, there's much more going on here than just that.

Media of every kind are portals to observational consciousness. And while we don't normally think of them in that way, that is what they are. Using and understanding media are two very different things. No one saw or understood this better than Marshall McLuhan and no one articulated the importance of understanding media better than he did in his seminal book, "*Understanding Media—The Extensions of Man*" published in 1964.

He separated different kinds of media into two broad categories, hot and cool. One reason he did that was to illustrate how differently we are affected by the media we choose. Hot and cool media make different demands on us and our innate abilities and, as we learn to use them, we facilitate the flow of observational consciousness in different ways because we access them via different realms of perceptual consciousness in different ways.

Radio is a hot medium because, like other hot media, it extends one perceptual realm—*hearing*—in high definition; with high definition defined as the state of being well filled with data. Because they extend one perceptual realm, hot media are low in user participation. They don't leave very much to be filled in by the user or, in the case of radio, the listener.

Television is a cool medium. It blends the visual and aural realms of perceptual consciousness but, unlike radio, it demands very little in the way of participation. When we watch television we're viewing a constantly changing orderly mosaic of pixels dance on a two-dimensional screen. Not much effort is needed to figure out what's going on. Watching television is non-participatory. To a large extent, the experience just *happens* to us which is opposite from the experience we have when we listen to the radio.

I still remember the occasion which helped me to not only truly understand but to actually *grok* what McLuhan meant when he described the difference between a hot medium like radio versus a cool medium like television.

It was the early 70s and I was driving to work, listening to the radio. The announcer said; "This morning we're going to do something really spectacular. We're going to drain the Chicago River, fill it with hot chocolate and drop a twenty ton marshmallow into it."

I remember smiling and thinking to myself, yeah right.

Then the announcer said, "We're beginning to open the drain now." In the background I heard sound effects depicting someone turning a large wheel connected to a mechanism which would open the drain in the bottom of the river. All the while, the announcer described what was happening, saying things like, "We're having trouble getting the drain to open but, wait a minute, and yes the wheel is moving. And now, the drain is open." With that, the sound effects changed. I heard sloshing and gurgling noises as the announcer described a huge vortex forming in the water above the open drain. This was followed by more sloshing and gurgling, with a final very loud whooshing sound as the narrator announced that all the water from the river had been drained. Then he said, "We're closing the drain now." And I heard more metallic clanking and grinding sounds as he described watching the large metal drain close.

"And now for the hot chocolate;" the announcer said. "Open the spigot."

I heard more metallic grinding and cranking sounds, followed by a loud rushing sound like a waterfall. "Oh man, I wish you could all see this," he said. "The river is filling with very hot chocolate. Steam is rising from its surface and there are bubbles everywhere. And the smell; it's just magnificent."

Then I heard the sound of a helicopter and the announcer said; "Here comes the marshmallow. The chopper is about 200 feet above the river and the pilot is getting ready to release the marshmallow. Here it comes." And with that, there was a snapping sound followed by a huge splash.

"We've done it, ladies and gentlemen. Drained the Chicago River, filled it with steaming hot chocolate and dropped a twenty ton marshmallow right into the middle."

What he said next is what drove McLuhan's point about the difference between hot and cool media home for me. "Let's see you do *that* on television."

In the early 70s we didn't have the CGI capabilities in movies or on television that we've all become accustomed to now. Visual special effects were severely limited and nobody could have visually pulled off the stunt the radio announcer had described. The only way to do it was with a hot medium like radio. Television was out of the question.

In thinking about this, you might be tempted to conclude that like television, film too is a cool medium; but it's just the opposite. Film is a hot medium.

> The mode of the TV image has nothing in common with film or photo, except that it offers also a nonverbal *gestalt* or posture of forms. With TV, the viewer is the screen. He is bombarded with light impulses that James Joyce called the "Charge of the Light Brigade" that imbues his "soulskin with subconscious inklings." The TV image is visually low in data. The TV image is not a *still* shot. It is not photo in any sense but a ceaselessly forming contour of things limned by the scanning-finger. The resulting plastic contour appears by light *through*, not light *on*, and the image so formed has the quality of sculpture and icon rather than of picture. The TV image offers some three million dots per second to the receiver. From these he accepts only a few dozen each instant, from which to make an image. The film image offers many more millions of data per second, and the viewer does not have to

> make the same drastic reduction of items to form his impression. He tends instead to accept the full image as a package deal. In contrast, the viewer of the TV mosaic, with technical control of the image, unconsciously reconfigures the dots into an abstract work of art on the pattern of a Seurat or Rouault. If anybody were to ask whether all this would change if technology stepped up the character of the TV image to movie data level, one could only counter by inquiring, "Could we alter a cartoon by adding details of perspective and light and shade?" The answer is "Yes," only it would then no longer be a cartoon. Nor would "improved" TV be television. The TV image is *now* a mosaic mesh of light and dark spots which a movie shot never is, even when the quality of the movie image is very poor.[21]

A movie is a series of *still shots* and each one is well filled with data. Each frame is made up of hundreds of millions of silver halide crystals comprising a fully formed image, which is why we accept viewing a film as what McLuhan described as a package deal.

The same is not true for television. Even with HDTV, the image resolution doesn't compete with the resolution available with film. And, while it's true both film and television present us with moving images, the TV image is a constantly changing mosaic of pixels whereas the film image is a serially presented series of very high definition still shots shown at a frame rate high enough to create the impression of fluid motion.

Film is hot. Television is cool.

The effect of television on our sensorium is very different from the effect of film or radio. But unless we come to understand not only how but also why this is so, most of us assume that watching a movie in a theater is essentially the same experience as watching television. It's not the same experience because, as McLuhan proclaimed, *the medium is the message.*

The media we choose shape who and what we are by presenting our sensorium with a gestalt peculiar to the technical dynamics which underlie how they operate. The reason they shape us is they shape observational consciousness. The media we choose have a direct, ongoing and lasting impact on who we are and what we are becoming. The medium is, indeed, the message.

When we experience electronic images, whether visual, audio or a combination of visual and audio, we're accessing the same realms of consciousness we use for direct perceptual experience. When I'm watching and listening to the squirrel in my back yard, I'm using the same aural and visual realms I use when I watch television. The important difference is the television image is by definition observational because it can be viewed, from the same perspective, by everyone else—simultaneously. It is the medium that makes the images observational. Observational consciousness is the mechanism, through which, the medium *becomes* the message.

We use electrons to create the visual and aural components of television images, or, for that matter, any kind of electronic images. We compress space, time and mass, in the arrangement of transmitted electrons and expand them onto the screen and speaker of the television set and, by doing that, we re-create the context of the images we transmit. What's going on, when we do this, is comparable to the chemo tactile stage of multicellular communities. In a chemo tactile organism, like a plant, each individual cell is capable of experiencing what the organism as a whole can experience.

In the same sense, observational consciousness is encoded in television images in much the same way it is encoded when we perceive perceptual objects like squirrels or coffee cups.

With the television image, *everyone* can perceive the information being presented. With the squirrel or coffee cup, the experience is limited to the *individual* making the observation.

In the future, we may come to experience things we now experience in observational consciousness, via television and other media, that is—as a whole—in a realm or realms of consciousness that we do not experience as individuals. If that happens a new and different structure of consciousness would replace the one we have now and to which we have become accustomed.

As we continue to use—adopt and adapt to—the ever-growing number of devices which expand and amplify observational consciousness, we spend more time using those devices and observational experience becomes increasingly more important than *direct* perceptual experience. We come to rely more on what we see and hear using these devices to gain knowledge about what's happening in the world and around us than we do on direct perceptual experience.

We see and hear about what's happening in China, Iraq, Iran and Afghanistan as well as how interest rates are being impacted by what's going on and we make decisions based on what we see and hear using these devices.

What we know and how we react to what's happening is becoming much less a matter of direct individual perceptual experience and more dependent on what we see on the screens and hear through the speakers and ear buds of our devices. Today we all do things based on *common* experience to a much larger extent than was the case for our parents and grandparents.

Twenty years ago, the space in which we did things was mostly perceptual. Now, that space is rapidly becoming electronic and thus observational.

A sense of place

In 1990, our sense of *place* was more important and intense than it is now. In order to do something, we had to go somewhere. The perceptual space we occupied was central to who we were and what we were able to do. The importance of the PC was beginning to ramp up as businesses, large and small, began to adopt and adapt to the technology. Spreadsheets, word processing and database programs were the lure that drew business to the PC. As personal computers became more powerful and less expensive, the software available became easier to use, better and more useful. We moved from the command line as an interface to the graphical user interface or GUI and Windows became the metaphor for using the PC and getting it to do what we needed done. And, even through all that, our sense of place and its importance remained strong.

By the turn of the century, what we could do with the PC expanded as we collectively discovered the World Wide Web and the internet. Our focus regarding, not only what we could do with the PC but what it was for began to change and, along with that, so too did the importance of place.

With a PC, printer and fax modem it became possible for an entrepreneur, working from home, to *look like* a much larger and more sophisticated business to vendors and clients. Word processing programs came with templates which could be used to create virtual company stationery with fancy letterheads that, when transmitted electronically or printed on blank paper, looked every bit as good as the letterheads used by Fortune 500 companies.

For doing business, the importance of our sense of place began to diminish even more as we continued to take advantage of what could be done using a PC and the internet.

Today, with the availability of smart phones, laptops and net books, and the availability of always-on wireless connectivity to the internet, our sense of place or where we are, is becoming even less important regarding what we are able to do.

With a Kindle electronic book from Amazon.com, I can order a copy of the latest best seller with the push of a button and have the book in under a minute. And for those of us who prefer hard copy books we can order them from Amazon.com from our smart phones, laptops and net books with a few clicks and have them delivered to our door. With the advent of the iPad from Apple, industry pundits are predicting that using that device and others like it, we will, at long last, be willing to pay for content like newspapers and magazines with the added advantage of having the content enhanced and augmented by the built-in video and audio capabilities inherent in the design of tablet computing devices.

What's happening, without our even being aware of it, is the space that our bodies appear to occupy is becoming rapidly less important to who we are and what we can do because the space, in which we do things, is fast becoming less perceptual and significantly more observational—more electronic.

Like it or not, whether we want to or not, we have begun to transcend perceptual space. The importance of *where* we are is quickly beginning to become vanishingly small.

As important and significant as transcendence of perceptual space is, in terms of who we are and what we are becoming, transcendence of the tactile realm of consciousness will be even more significant and will, ultimately, exert an even stronger influence on who we are, what we can do, what we are becoming and how we relate to each other and the planet.

> From the evolution of the cell membrane to the age of the internet, the axis of doing has been the tactile realm, but as potential perception becomes as real as actual—as I feel your pain as much as I feel my own—the body will be less attached to self and doing.[22]

Forty years ago, in 1970, if you wanted to see a movie you had to go to the theater. Today with big screen TVs surround sound and on demand services from cable companies and Netflix, you can choose a movie from a vast selection of relatively recent releases by touching a few buttons on your remote control. And, you can pause it, rewind it or replay it with little more effort than thinking about what it is you want to do.

For watching movies, reading books, buying goods and services of almost every imaginable kind, the axis of doing is no longer embedded in and tied to the tactile realm of consciousness. You don't have to *go* anywhere. You just do it. The body truly is becoming less attached to self and doing. This evolutionary process, in which we find ourselves embedded, is not entirely conscious. Put differently, even though we are participants in the process we are not entirely mindful of what the process is or how we are being impacted by what is happening. We build televisions, computers and cell phones and, as we adopt and adapt to the new technologies, we extend innate capabilities out into the environment.

The lever principle allows us to extend and amplify our musculature. The wheel as an extension of our feet allows us to move our belongings and ourselves much further and faster than was possible without it. Steam shovels extend and amplify our musculature even further than the lever principle alone. The industrial revolution built upon these advances and amplified and extended into the environment what we could do and what we could build.

With electronic technology, including the telegraph, the telephone, radio and television, we began externalizing and extending, into the environment our nervous systems. As we continue to adopt and adapt to computers, we are extending our brains into the environment; externalizing and amplifying our ability to think.

As we continue on this path, embedded in the process, we are drawn along by it. We are not mindfully attempting to become a larger self but, ultimately, that is what's happening. The technology truly does enhance our ability to feel each other's pain to a much greater extent than was possible without it. Whether we like it or not our capacity for empathy is expanding along with our increasing dependence on and the importance of the observational realm of consciousness.

Potential perception is becoming every bit as important as actual perception.

While surrounded by and embedded in the rapid emergence of newer, more powerful, devices which extend and amplify observational consciousness, we get the impression that the devices almost seem to build themselves. We assume they have been invented to do things and while, on one level, that is certainly true; on another level, to a much larger extent than was ever true or possible before, they have become part of whom we are.

I've heard and read comments by individuals suggesting life would be easier and better without all this constant on availability of communication made possible by cell phones and smart phones. And, while, from a certain narrow perspective, I understand and commiserate with the opinion, in a broader sense, abandoning our e books, iPads, cell phones and smart phones would be self destructive. Whether we like it or not; whether we want to or not, we have to keep them. But we also have to become aware of how tightly they hold us together. One important consequence of how tightly we are being held together by our new devices is the extent to which they have amplified how strong we are. We are rapidly becoming stronger than we know how to be. This is a dangerous situation because we are becoming too strong to live divided in a place that is so small.

As the importance of our individual sense of place diminishes, and our collective strength increases the need for and importance of empathy becomes significantly more intense. The path to dealing effectively and safely with the danger associated with our ever-growing strength is, once again, learning to do *collectively* as we would be done to. That will be—as it must be—a conscious process.

The speed and intensity of change as it affects who we are and who we are becoming is difficult to grasp because the average lifetime, compared to the broad sweep of recorded history, is relatively brief. To get an idea of what this means and how important it is consider this statement from the book *Future Shock* written in 1970 by Alvin Toffler.

> [It has been observed]... that if the last 50,000 years of man's existence were divided into lifetimes of approximately sixty two years each, there have been about 800 such lifetimes. Of these 800, fully 650 were spent in caves. Only during the last seventy lifetimes has it been possible to communi-

> cate effectively from one lifetime to another—as writing made it possible to do. Only during the last six lifetimes did masses of men ever see a printed word. Only during the last four has it been possible to measure time with any precision. Only in the last two has anyone anywhere used an electric motor. And the overwhelming majority of all material goods we use in daily life today have been developed within the present, the 800th, lifetime.[23]

What makes that statement so stunning is its focus on how technology has changed our lives and how we lived and related to each other before adopting and adapting to technologies that, for most people alive today, have been around since before they were born. What makes it as amusing as it is astounding is, in 1970 there were no personal computers.

Science is the primary driver behind the growth of observational consciousness and the growth of science is what underlies technology, the new devices that are emerging, and the influence they are having on how rapidly observational consciousness is expanding.

Even before the emergence of electronics and mass communication, a scientific fact was, by definition, an observational experience. What makes scientific facts so valuable is they represent not only what you experience or I experience but what you or I or *anyone* can experience at any time under the same circumstances. Scientists continuously create what is known by doggedly insisting that the experiments they conduct must be repeatable and results must agree. They accomplish this by agreeing, always, to follow the principles of the scientific method and because they do that, their ongoing efforts produce a permanent and reliable record of potential perception.

For those unfamiliar with science and how it works, the importance of all this is difficult to understand.

In addition to producing a permanent and reliable record of potential perception scientists also simultaneously create a dynamic construction of consciousness more important and more profound than language. Their efforts are—by definition—universally human.

A scientific fact is incontrovertible because, if it were not, it wouldn't be science. That's the reason a scientific fact transcends the individual and collective perspectives of our separate selves and, in so doing, makes that separation vanish. For this reason, scientific facts have an apparent *objective* reality that does not depend on conscious experience.

> We always see them, no matter where or who we are. Even controversies surrounding scientific authenticity reveal the underlying structure of observational consciousness: Each side in a scientific controversy will demonstrate, through experiment, that a particular fact is or is not a potential perception.[24]

The extent to which what scientists do is entangled in observational consciousness, is not intuitively obvious until we become aware of the existence of observational consciousness, how it works, and why it's important. Observational consciousness existed long before we began to embrace even primitive technologies let alone adopt and adapt to them en masse.

The myth and metaphysic emergent from the dimensional structure of consciousness illuminates both the existence and importance of the sixth realm of consciousness—observational consciousness. The spike we are experiencing in the growth of observational consciousness is, arguably, the most significant event of our present era. As we participate in that growth, wittingly and unwittingly, we are simultaneously participating in and witnessing the evolution of consciousness to a higher level. But, even so, even with our televisions, personal computers, smart phones and iPads, observational consciousness is not as vivid, well formed, or highly structured as direct perceptual experience. It's getting there but it isn't there yet; and it won't be until we develop technology similar to the holodeck.

Less well-structured and vivid still, than observational consciousness are thought, dreams, illusions, religious experience, and myth. We are gradually coming to understand that while these things are more difficult to experience, they are every bit as real as observational consciousness and direct perception.

The dimensional structure of consciousness provides a framework for understanding that enables us to appreciate how the universe works. To understand that when we perceive an object like a coffee cup or lamp, while it's comfortable and easy to assume that object exists "out there" because we can see it and touch it, the existence of any object can also be understood as our experience of that object and that, absent our experience of it, the thing we perceive has no objective existence—out there or anywhere else. The ongoing and rapid expansion of observational consciousness and its evolution to a higher level of consciousness will help us to understand and ultimately come to grips with the deep implications embedded in the dimensional structure of consciousness. Ultimately we will be able to apprehend the emergence of a new paradigm that more fully reveals what consciousness is and how it works.

Some thoughts about thinking

Most people know what thinking is because most people have experienced it. A question naturally arises when considering what thinking is—when we're thinking are we really *doing* anything? It's difficult to say because doing is, more often than not, associated with events that occur in dimensions—things we can perceive through vision, hearing, smell, taste and touch.

If you see someone sitting quietly in a chair with his eyes closed you can make several assumptions about what he is doing. He could be taking a nap or meditating or he could be thinking. But, just looking at him doesn't really provide any information about what's going on.

If his eyes are open and his facial expression gives the impression that he is deep in thought that may or may not be what's happening. He may be deep in thought but, it is also entirely possible that he's simply trying to create the impression that he's deep in thought. As Samuel Avery observes in *Transcendence of the Western Mind*, "Thinking is usually something you do when you're not doing anything,"

What can we really say about thinking?

- Thinking is difficult
- Thinking is easy
- Thinking is fun
- Thinking is a waste of time
- Thinking is for nerds
- Thinking is dangerous
- Thinking is painful
- Thinking is pleasurable

- Thinking is what we do when we don't have anything better to do

Thinking is individual and *internal* in the same sense that perceptions are individual and internal. My perception of the color red is mine alone and happens in my head. The same is true of my thoughts and everyone else's.

Thinking is a somewhat slippery concept. This particular thought I'm having is no longer mine alone if you are reading these words because, by typing it on my word processor [*doing*], the thought became dimensionalized and coordinated with seeing. I stuffed it into space and time and mass and, when I did that, it became done. And, now that it is done, it can be interchanged with space, and time, and mass so that I can see it as well as anybody else. This simple example of how a thought becomes dimensionalized through doing by cramming it into space and time and mass leaves out a vast category of images that cannot be crammed into space and time and mass because they simply don't fit.

> We tend to forget about the stuff that does not fit in space or mass because it cannot be done, and we forget about the stuff that does not fit in time because that is what forgetting is. But as soon as it fits in the dimensions, "I" comes with it. As soon as it is actualized within a potential, there are many more things that it is not, and "I" only sees what is and not what is not. It is the "I" that sees the actual, so "I" goes in the dimensions along with the seeing. All that empty space around it is just potential experience that "I" never sees. Empty space has to go along with "I" so that it does not see everything. If "I" were to see everything, there would be no potential experience and no dimensions and no use for "I."
>
> Thinking divides being and never gets at the thing itself, if there is such a thing. It is never more than the best I can do. It is images and words that are images that stand for other images churning and grinding and bearing on one another. The words are sounds that are not themselves the images they mean. I say and write what I think and I think of you thinking the same thing, but I know that does not happen. I know that nobody else sees it. It is the structure "anyone else" that is real—the structure that makes you hear me say something you recognize and makes you nod. That is what I want to find—the structure that makes you nod.[25]

There is an awful lot of information crammed into that passage with a tight focus on thinking but, not so much about what thinking is but rather more about what thinking does and what it cannot do. It cannot, for example, dimensionalize dreams or religious experiences by cramming them into space and mass. They don't fit in space and mass.

Thinking is the process that creates the images and words that are images that stand for other images. Thinking helps us to extract meaning from the thoughts we experience and shape what is available in the churning and grinding.

Even after we have extracted meaning from our thoughts we realize the words which represent the meaning are sounds that are not themselves the images they mean. But, even after all that, if we have been careful and disciplined enough in our thinking, when we say or write what we thought, "anyone else" can hear or read what we have said or written, understand it and nod. That structure "anyone else" is the nexus between thinking and understanding; the structure Avery wants to find.

Meditation and thinking are very different. Thinking requires me or "I" or self and, ultimately, becomes the things that "I" see and do. But, it is more than just that. Thinking, in a very real and immediate sense, creates me. "Cogito ergo sum" or "I think therefore I am." Thinking is a way of *being* in the world. When what I think becomes dimensionalized that's when it becomes what I see and do; and, if I do it just right, under the right circumstances other observers can see what I thought. Thinking is the process or mechanism we use to dimensionalize some small portion of what our thoughts reveal. But thinking does not allow us access to the chaos.

If you want to *see* the chaos as it churns and grinds and spews images and ideas that float and twist and fade out of sight and, mostly, out of time, you need to find a way to transcend thinking with its focus on straining the chaos through space and time. You need to find a method to detach from the self that's doing the thinking and straining what it is doing through space and time. You need to detach from the world because the world is space and time and labels. Meditation is one way to do that and, if you succeed, you are no longer you—*you* no longer *are.* And, while success allows you to glimpse the chaos you cannot remember what you glimpsed because you were not *you* when it happened.

Glimpsing the chaos requires the reverse mechanism associated with "I think therefore I am," a mechanism that allows you to detach from "I" and the world and *see* in a way that has nothing to do with dimensions or perceptions or the senses. But, for most of us, most of the time, thinking is the best we can do and has gotten us to where we are. Descartes understood this and so do we. We think therefore we are.

> Thoughts arise and fall, some in dimensions, some not, some trying to get in. That is all there is. [26]

What follows are some thoughts that, when I first encountered them, made me nod. And each time I read them, I nod again. They are from *the Tao Te Ching* written by Lau Tzu 2500 years ago.

> We join thirty spokes
> to the hub of a wheel,
> yet it is the center hole
> that drives the chariot.
>
> we shape clay to
> birth a vessel,
> yet it's the hollow within
> that makes it useful.
>
> We chisel doors and windows
> to construct a room,

yet it is the inner space that makes it livable.

Thus do we
create what is
to use what is not.[27]

That which we look at
but cannot be seen is the invisible.

That which we listen to
but cannot hear is the inaudible.

That which we reach for
but cannot grasp is the intangible

beyond reason, these three merge,
contradicting experience.

Their rising side isn't bright
Their setting side isn't dark.

Sense-less, unnamable, they return
to the realms of nothingness.

Form without form,
image without image,
indefinable, ineluctable, elusive.

Confronting them, you see no beginning.
Following them, you see no end.

Yet, riding the plowless plow
can seed the timeless Tao,
harvesting the secret
transcendence of the Now.[28]

Long before we understood the basic principles of physics or had access to any of the technologies we were able to create once we understood the fundamental principles, observational consciousness existed but it was harder to access.

Lao Tzu was able 2500 years ago to write down his thoughts about how he believed the universe works. And, because language existed then and his words were preserved, we are now able to access those thoughts through translations. When we read what he had to say the dimensionalized thoughts make us nod.

When I read what he had to say, I'm struck with a sense of how much he struggled to understand how the universe worked and, even more, with how difficult it was for him to put into words what he was able to *see* regarding how the universe works and how hard it is to understand what's going on. And that is why—2500 years later—the structure embedded in what he had to say still makes us nod.

TIME

To help us make sense of our everyday experiences, in and of the world, we make several assumptions about time.

- Time is real
- Time moves or flows
- Time has a recognizable direction or arrow
- Time's arrow points from present to future
- Time exists as a dimension as in the space–time continuum

All this is well and good. It helps us to make sense of our everyday experiences and provides a mechanism for explaining what we observe. Our assumptions about time, its existence, and the way it works hold up fairly well until we begin to examine them carefully.

Newton believed that time flows uniformly throughout the universe and acts as a background, against which, events unfold. Put differently, he believed the universe has a kind of *master clock* that simply keeps on ticking.

When we think about time:

> We have a deep intuition that the future is open until it becomes present and that the past is fixed. As time flows, this structure of fixed past, immediate present and open future gets carried forward in time. This structure is built into our language, thought and behavior. How we live our lives hangs on it.
>
> Yet as natural as this way of thinking is, you will not find it reflected in science. The equations of physics do not tell us which events are occurring right now—they are like a map without the "you are here" symbol. The present moment does not exist in them, and therefore neither does time.

> Additionally, Albert Einstein's theories of relativity suggest not only that there is no single special present but also that all moments are equally real... Fundamentally, the future is no more open than the past.
>
> The gap between the scientific understanding of time and our everyday understanding of time has troubled thinkers throughout history. It has widened as physicists have gradually stripped time of most of the attributes we commonly ascribe to it. Now the rift between the time of physics and the time of experience is reaching its logical conclusion, for many in theoretical physics have come to believe that time fundamentally does not even exist.[29]

This quote is from an article in the June 2010 issue of *Scientific American* entitled "Is Time an Illusion?" by Craig Callender. It identifies the extent to which things begin to get dicey at the nexus between our commonsense understanding of time and what physicists discover when they examine time using mathematics, equations and experiments. For them, time disappears.

As I pointed out earlier, even though most people believe time exists, no one has been able to *detect* time in the same sense that we can detect other physical phenomena like light, heat or mass.

So what is it then, that provides us with the deep sense that time not only exists but also flows? One answer to that question is change. We get older. Our hair gets grey, seasons come and go and we are able to observe the changes that occur over time. Given enough time, every object in the universe will eventually scatter its atoms to the wind. And yet, when we search for physical evidence of the existence of time using mathematics, the discipline of science, and the scientific method, we consistently come up empty handed.

It seems clear that, as scientists intensify their focus on finding and defining time, their efforts increasingly seem to indicate that time, as we have come to understand it, in our everyday experience—doesn't exist. For most people, that's just crazy. Our hair does turn grey, we do get older and, eventually, we die. Without including time, as we understand it, at first blush, it appears impossible to explain the seasons, our graying hair and our mortality. But, if we think about this carefully, it turns out that it is possible to explain these seemingly time dependent phenomena without resorting to time. Change can be described without depending on time.

When we think about a season change like from spring to summer, we normally relate that change to the passage of time associated with how long it takes for the earth to complete that portion of its transit around the sun. But we don't have to do that. We could just as easily relate it to how many times a metronome, sitting on a piano in my living room, ticks between spring and summer. If we choose to use the metronome, then, what we are doing is describing the correlation between two physical objects, the earth as it orbits the sun and the ticking metronome. And if we imagine the metronome is electronic and has a counter that keeps track of the ticks and displays how many ticks have occurred since we started it then we have described the change from spring to summer with no need for or reference to a global time. A ticking metronome is not time; it is a metronome.

Instead of saying a car accelerates at a rate of fifty meters per second we could describe it in terms of how tall a bamboo plant grows in my backyard. And, for something like the graying of my hair, I could just as easily relate it to changes that occur in a glacier. In all these examples we've managed to describe change by correlating changes between two physical objects thus eliminating the need for time. In all three examples time becomes redundant because we are successfully describing change without it.

Now, while it's possible to describe change in terms of correlations between two physical objects, it becomes obvious that keeping track of all the changes that constantly occur around us by correlating changes in physical objects would be difficult to do and keep track of. So instead we have organized this vast network of correlations and defined something we "call" time. Then we relate everything to it which nicely eliminates the trouble associated with trying to keep track of all those correlations between physical objects.

That's essentially what physicists do. They describe and summarize what is happening in the universe in terms of physical laws that play out in time. But, when they try to find time, in their mathematics and equations, it vanishes. So one conclusion from all this is, that while it is convenient to use time to describe change and what happens when we conduct experiments, it is a mistake to assume that time really exists.

If you think that all this is a stretch, think about money. Money is just something we invented as a placeholder. Without money, every time I wanted get some beer, I'd have to figure out what I could trade for it. I'd have to barter for everything I needed and so would everybody else. Money eliminates the need to barter and makes the exchange of goods and services much easier and more convenient but, just as with time, it is a mistake to assume that money *really* exists. It is nothing more than a placeholder and the only reason it has any value is because we say so and agree to rules about how it's produced, denominated and kept track of. As pointed out earlier, for most of us, most of the time, money is nothing more than ones and zeros on the surface of a hard disk on a server in *the cloud*.

It is entirely possible and, in fact, probable that time, as we have come to understand it, on the macroscopic level of existence we inhabit is an emergent phenomenon. We apprehend it on the macroscopic level but, on the quantum level it vanishes.

We are all familiar with emergent phenomena or properties of physical objects or systems. The wetness of water is just one example. Water is wet. But, when we examine water we discover that it is made of molecules composed of one oxygen and two hydrogen atoms. Hydrogen is a gas and so is oxygen. Neither of them is wet. But, when hydrogen and oxygen combine, in just the right way, water and wetness emerge on the macroscopic level of existence. On the quantum level wetness, like time, vanishes.

On the macroscopic level, a boulder is solid and hard. But, when we examine it carefully, we discover that it is composed of atoms which are composed of protons and neutrons in their nuclei with electrons orbiting those nuclei. And when we examine those atoms we find that they are mostly empty space. On the quantum level solidity and hardness vanish. Solid objects *appear* solid in the middle latitude of reality where we exist. Solidity, like wetness is an emergent property and, it is important to keep in mind that this is not just a guess. It is something we *know* because we have carefully examined matter and scientifically proven that it is composed of atoms. Likewise, we have carefully examined atoms and proven that they are composed of protons, neutrons and electrons.

Commonsense provides suitable explanations for what we perceive using vision, touch, taste, smell and hearing but, when we examine those same perceptions using science and the discipline required by the scientific method, more often than not, our commonsense understanding of what's really going on, crumbles and falls apart. To a very large extent, this is what's happening with our understanding of time. Commonsense tells us it really exists. Careful and disciplined scientific examination of time, however, tells us it does not.

What's truly interesting and surprising about all this is, after all is said and done, what ends up changing is our *understanding* of what we are experiencing. Our actual experience remains unchanged.

There was a time when we believed the earth was flat. And now, even though we know it is a very large sphere, when we look around, it still looks flat. The last time I was on a cruise, while at sea I looked out from the balcony of my stateroom and noticed that, even though I know the earth is a sphere the sea still appeared to be flat. Then, I looked at the horizon and noticed that it described a curved line rather than a straight horizontal line. Given the right circumstances, the earth's spherical shape reveals itself even if you are still on the surface.

When we came to understand that the earth was not the center of the universe and that, when we saw the sun rising in the morning and setting in the evening it was the earth that was moving and not the sun, what we saw in the morning and evening sky *looked* the same. What changed was our *understanding* of what we were seeing.

As scientists continue to examine time I believe they will eventually reach consensus and agree that it doesn't really exist just as they did with their understanding of the shape of the earth and its position in the cosmos relative to the other heavenly bodies they could see. If that happens, just as before, when we look around, the world will *look* the same as it does now. The only thing that will have changed will be our *understanding* of time.

If time doesn't exist, other than change, what is it that creates the impression that it exists and flows in an identifiable direction. Newton believed there is a master clock that ticks equably providing the background of time against which events, in the universe, unfold. But, if Newton was wrong, as physics experiments seem to strongly indicate, what is it, other than being able to perceive change that creates the impression of an omnipresent ticking associated with our experiences on the macroscopic level of existence?

Well, for one thing, we tick. We have a heartbeat. For as long as we are alive, that constant lubdub going on, in our chests, acts as an omnipresent internal clock. Beyond that, when we examine the universe we find that everything is always and only in exquisite motion and, even though we are sometimes confronted with what *looks like* stillness on the macroscopic level, we know that stillness does not exist. The stillness we sometimes think we see is an illusion.

Electromagnetic radiation also ticks. It has wavelength and frequency that when multiplied always yields the universal constant c. And, even when we examine matter, we find that it too ticks in the sense that electrons are furiously orbiting atomic nuclei and the nuclei themselves are furiously exchanging particles to maintain nuclear integrity.

If we look carefully we find that everything ticks. But, in all this, I believe it is important to keep in mind that ticking of any sort is not time for the same reason that a metronome or a clock is not time. Time is not the same thing as ticking.

A ticking clock is not *measuring* anything. The number of ticks a clock makes during one second is arbitrary as is the length of a second. All these phenomena are created to model the earth as it spins on its axis.

Now is all we ever really have.

Experiences outside dimensions

Our Western mindset is fashioned on principles laid down by Rene Descartes. The Cartesian worldview cautions us to be wary about experiences that are not objective. Colors, smells, tastes, feelings and intuitions are not to be trusted. The way to remain on solid ground is to focus on matter and material things.

It took hundreds of years for the Western mind to become fully inculcated with Cartesian duality and, by the middle of the twentieth century, the transformation had become complete. We focused hard on science and the scientific method and were handsomely rewarded for our efforts with wonder drugs, radar, atomic energy, x-rays, atomic bombs, hydrogen bombs, the telegraph, the telephone, radio, television, computers and expanded understanding of how the universe works.

The harder we focused, the more we discovered. Science and technology became a seemingly endless cornucopia of innovation and development. We locked ourselves in and our success was spectacular. Continued success convinced many that any imaginable problem could eventually be made to yield to careful and rigorous objective scientific analysis. And so we've come to where we are now. Our mastery of technology is magnificent.

One measure of technological competence is the ability to manipulate energy. Learning to control and use fire is an example of early technological competence. We have come far since succeeding in our quest for and eventual ability to control and use fire. Today, our ability to manipulate energy allows us to move individual atoms with enough precision to form readable letters.

Our ability to exert exquisite control over electrons is what makes our computers do what we tell them to do and what allows us to communicate using cell phones. In learning to master science and technology we have gained much but, in doing that, we have also lost much.

We have locked ourselves into objective reality and learned to see the world as existing only in terms of what we and other observers can see, touch, hear, taste or smell. We think of ourselves as our bodies. But, that is a mistake. We are much more than just our bodies. R. Buckminster Fuller understood this implicitly.

"The Omnidirectional TV Set"

> Children looking at TV today look at it quite differently from the way it was to the first generation of TV adults. It begins to be very much a part of the child's life, and he tends to accredit it the way adults accredit what they get from their eyes. When children are looking at a baseball game, they are right there in the field. All of our vision operates as an omnidirectional TV set, and there is no way to escape it. That is all we have ever lived in. We have all been in omnidirectional TV sets all our lives, and we have gotten so accustomed to the reliability of the information that we have, in effect, projected ourselves into the field. We may insist that we see each other out in the field. But all vision actually operates inside the brain in organic, neuron-transistored TV sets.
>
> We have all heard people describe other people, in a derogatory way, as being "full of imagination." The fact is that if you are not full of imagination, you are not very sane. All we do is deal in brain images. We traffic in the memory sets, the TV sets, the recall sets, and certain incoming sets. When you say that you see me or you say "I see you," or "I touch you," I am confining information about you to the "tactile you." If I had never had a tactile experience (which could easily be if I were paralyzed at conception), "you" might be only where I smell you. "You" would have only the smellable identity that we have for our dogs. You would be as big as you smell. Then, if I had never smelled, tasted, nor experienced tactile sensing, you would be strictly the *hearable you*.
>
> What is really important, however, about you or me is the *thinkable you* or the *thinkable me*, the abstract metaphysical you or me, what we have done with these images, the relatedness we have found, what communications we have made with one another. We begin to realize that the dimensions of the *thinkable you* are phenomenal, when you hear Mozart on the radio, that is, the metaphysical—only intellectually identifiable—eternal Mozart who will always be there to anyone who hears his music. When we say "atom" or think "atom" we are intellect-to-intellect with livingly thinkable Democritus, who first conceived and named the invisible phenomenon "atom." Were exclusively tactile Democritus to be sitting next to you, surely you would not recognize him nor accredit him as you do the only-thinkable Democritus and what he thought about the atom. You say to me: "I see you sitting there." And all you see is a little of my pink face and hands and my shoes and clothing, and you can't see *me*, which is entirely the thinking, abstract, metaphysical me. It becomes shocking to think that we recognize

> one another only as the touchable, nonthinking biological organism and its clothed ensemble.
>
> Reconsidered in these significant identification terms, there is quite a different significance in what we term "dead" as a strictly tactile "thing," in contrast to the exclusively "thinking" you or me. We can put the touchable things in the ground, but we can't put the thinking and thinkable you in the ground. The fact that I see you only as the touchable you keeps shocking me. The baby's spontaneous touching becomes the dominant sense measure, wherefore we insist on measuring the inches or the feet. We talk this way even though these are not the right increments. My exclusively tactile seeing inadequacy becomes a kind of warning, despite my only theoretical knowledge of the error of seeing you only as the touchable you. I keep spontaneously seeing the tactile living you. The tactile is very unreliable; it has little meaning. Though you know they are gentle, sweet children, when they put on Halloween monster masks they "look" like monsters. It was precisely in this manner that human beings came to err in identifying life only with the touchable, physical, which is exactly what life isn't."[30]

That passage is from what, at the time it was published, was R. Buckminster Fuller's magnum opus—*Synergetics*. Arthur C. Clarke called it "the distilled wisdom of a lifetime."

In re reading it, I was struck by the similarity of Fuller's description of the omnidirectional TV set to the way Samuel Avery describes vision and scrolling the photon screen.

Fuller also said, "No one has ever seen outside themselves." When you think about that statement carefully, it becomes impossible to deny the truth of it but even more so, in light of the dimensional structure of consciousness and how it works. When we look at any object, what we ultimately see is the result of photons touching photoreceptor cells in our retinas. Because photons move very fast, we are fooled into thinking that we see things instantaneously. We do not. If you are looking at these words right now, it *appears* that you see them instantaneously because it takes only about a billionth of a second for the photons from the page to reach your retinal cells. And, while it's certainly true that a billionth of a second is very fast it is not instantaneous.

When you look at yourself in the mirror, what you see is the *touchable* you. What you don't see is the metaphysical/spiritual only *thinkable* you. As long as you are alive and in possession of all your sensorial faculties, this mechanism is always on, always operational. And it applies to everything and everyone you perceive. What you experience in the perceptual realms happens in your head even though it appears to be happening *out there*. There is no *out there*. Light and its speed, c, is what enables and sustains the illusion. Space–time is in light rather than the other way around.

Fuller's understanding of what happens when we see something was syllogistic. And even though he did not know about or understand the dimensional structure of consciousness, he clearly understood how vision works and the importance of the difference between what we are able to perceive in the perceptual realms versus what we cannot perceive within the dimensions.

We have all had experiences outside the dimensions. Falling in love is a nondimensional experience. It's grounded in feelings and desire and has nothing to do with objectivity.

If you've had the good fortune to have owned or played with a kitten, you know, first hand, how delightful that experience is. Kittens and puppies generate delight which is nondimensional and non-objective.

Other common non-objective nondimensional experiences include hate, envy, lust, greed, shadenfreude, desire, sorrow, sadness, happiness, joy and bliss. These kinds of experiences are emotional but they are also spiritual and metaphysical. They are associated more directly with the *thinkable* you than they are with the *touchable* you. Dreams and hallucinations are also non-objective nondimensional experiences.

The thinkable/metaphysical you reaches out from the touchable/physical you and allows you to experience and appreciate nondimensional aspects of existence associated with other observers like kittens, other levels of consciousness as with mediation and dreams and with inanimate objects. These kinds of nondimensional experience are every bit as real as dimensional experience even though Descartes would not agree.

For many people, a car is not much more than a machine; something you use to get from point A to point B. For others, especially the subset of individuals known as car aficionados, gear heads or car guys a car is much more.

For car guys while it's true cars have wheels, tires, fenders, windshields and engines, they have much more; they have spirit and other metaphysical qualities which make them desirable in ways that have little or nothing to do with their utility or getting from place to place. For them, cars are iconic with deep and rich histories and they become objects which car guys lust after. The *thinkable* car becomes much more than just a machine made of steel, rubber, plastic, and glass. And, it is the *thinkable/metaphysical* car and its spirit and magic which appeals to and ignites the strong nondimensional experience car guys appreciate and enjoy.

Car aficionados are able to describe, in minute detail, the difference in performance and handling between say a BMW M3, or Porsche 911, and a Honda Civic. In objective terms they will tell you about suspension geometry, engine displacement, acceleration, valve trains, exhaust systems, torque and horsepower.

In nonobjective terms a car guy will describe how the car *feels* when approaching the apex of a turn, at speed, and the extent to which the car delivers information from the road to his hands and the seat of his pants. He will talk about the exhaust note and the extent to which it is pleasant, exciting, annoying or boring. He will describe the suspension geometry's ability to instill confidence in attacking turns on demanding track courses or winding roads. And most of what a car guy apprehends and appreciates about the cars that appeal is nondimensional.

Put someone who sees cars simply as appliances behind the wheel of a BMW M3 or Porsche 911 and their experience will be very different. They may notice that the cars feel different than the Honda Civic or other car they usually drive but the experience will not generate excitement or pleasure and they will remain mystified regarding how or why anyone would be willing to pay so much money for something that, from their perspective, isn't really that much different from what they normally drive.

Some examples of cars with style, strong sprit and strong metaphysical qualities or, what some call strong juju are Mustang, Corvette, BMW, Porsche, Bugatti, Lamborghini, Ferrari, Audi, Rolls Royce and Bentley.

What's interesting about all this is the metaphysical qualities, spirit and juju really do exist in the cars. It's tied to their history, performance capabilities, overall quality, and the care and expertise used to manufacture and market them. All these nondimensional attributes are obvious to car aficionados. Those who care little or nothing about cars cannot *see* them.

Even mundane objects have spirit but it is difficult to *see* because our Western/Cartesian mindset gets in the way. It is easier to *see* the spirit associated with iconic objects like the Empire State building or the Eiffel tower. The same is true of places like the Grand Canyon or the Amazon rain forest.

Cities too, have spirit and metaphysical characteristics. Chicago, Los Angeles, New York, Paris, Rome, Munich, London, Berlin, and Tokyo are iconic cities familiar to people all over the world. They have rich histories and a spiritual dimension that becomes stronger and easier to identify when you visit them. Their nondimensional/metaphysical attributes become even more pronounced if you have the opportunity to live in any of them for an extended period.

Our enjoyment of music is nondimensional. When we listen to music that moves us, that experience is similar to the one car aficionados have when they see or interact with a car they find appealing. And even though the experience is nondimensional, it can and often does bring intense pleasure.

Style is nondimensional. We all recognize it when we see it but its impact varies depending on what the observer brings to the experience. Style applies to objects, like cars or buildings as well as to words and ideas. And, here, again, car aficionados provide access to the intensity of nondimensional experiences through the style of the words they use in describing what it's like to interact with specific automobiles. To get an idea of what I mean, consider this from an article in the August 2010 issue of *Motor Trend* magazine written by Arthur St. Antoine entitled: "The Flyover—three very different very fast German supercars land in Vegas then take off." This article describes the experiences of three *Motor Trend* staffers as they drive an Audi R8 5.2, a Porsche 911 turbo and a Mercedes-Benz SLS AMG. The cars range in price from just over $150,000.00 to $203,500.00 or, as one writer put it, each one costs well over one hundred fifty large. These are the kinds of cars that provide intense nondimensional experiences for car guys; feelings of desire and the article describes the opportunity to experience driving cars that only a select few ever get to experience.

The group of three went to Las Vegas to drive the cars. They began their adventure by checking into the Bellagio Hotel and Casino for drinks in the Petrossian lounge. All three are bona fide gear heads and talented writers and, if you pay attention to how they say what they have to say, you will gain some understanding about how intense their nondimensional experiences were and the extent to which they were involved in and enjoyed what they were doing.

"I'm driving the Gullwing first tomorrow," says Ed Loh, leaning back, libation in hand. "I need seat time to decide whether the V-8 sounds like the engine room of the USS Nimitz, or KISS playing Wagner during a thunderstorm."

"Actually you're in the 911 Turbo," says Ron Kiino, wiping a dab of caviar from his lips. "I'm in the Audi R8, and Art's in the Gullwing. "I know this because—oh, sure, I'll have another cocktail, thanks—I know this because 563 horsepower is too much for you. Also, anything you want you cannot have."

"Fine, I'll gladly drive the Turbo," says Loh, "And you two will be looking at nothing but whaletail, because my Porsche can go from 0 to 60 in [voice rises to taunting falsetto] just 2.8 seconds. Plus with my skills I can...hey, is Art still here? All I see at his chair is an empty martini glass and a cloud of cigar smoke."

The trio had arranged to have a section of highway cordoned off by the highway patrol to do their testing.

Pay attention to the nondimensional aspects of what they say about their experience; how they describe what they're doing and how it feels as they do it.

"Our bright side couldn't be more brilliant. Downstairs, safely tucked out of reach for the night in the Bellagio garage, are three cars guaranteed to blow the fog (forecast for us: sure thing) from tomorrow's dawn. It's an intoxicating mix: the front engine/rear-drive Mercedes-Benz SLS AMG Gullwing, the mid engine/all-wheel-drive Audi R8 5.2, and the rear-engine/all-wheel-drive Porsche 911 Turbo. Engines: naturally aspirated V-8 and V-10, twin-turbo flat-six. Output 500 horsepower, minimum, Speed: yes, thank you."

"Anybody rich enough to own one of these cars would've slept in at least until two p.m." Loh's voice crackles over the radio as our Teutonic trio storms into the Nevada desert the next morning, Vegas quickly falling far behind, the sun still low on the horizon.

"Let's talk about how I was born to drive the Audi R8V-10," says Kiino. "Man, I love the smell of 145 mph in the morning."

"Now I know how it felt to be Slim Pickens at the end of 'Dr. Strangelove,'" I say from behind the wheel of the SLS. "Driving this thing is like riding an atom bomb—except with a really great stereo."

If you want specifications and track numbers, they're all here in the charts at the end of this feature. Read them carefully, then read them again. The first time, you might not believe your eyes (the slowest car here hits 60 mph in just 3.6 seconds). Our objective in gathering this trio, though, was to savor these machines as an owner might—flat footing them in the great wide open, screaming up and down vacant twisties, simply letting their wonderfulness flow over us for hours.

Already, one thing is clear: As a piece of styling drama, the Porsche ranks a distant third. "I love a good Q-ship," says Loh. "But the 911 Turbo goes too far. Simply disappears compared with the other two." At the Bellagio's valet stand, the attendant barely gave the Turbo's familiar 911 shape a second glance. In contrast, Kiino dubs the Mercedes SLS "a standout. From that long hood to its gullwing doors to its matte-silver paint...impossible not to get noticed." (Noticed? The thing is a rolling Paris Hilton—in Vegas, whenever the Gullwing rolled into view, awestruck passers-by whipped out pocket shooters and video cams or simply begged to sit in the driver's seat.) Loh has similar praise for the Audi: "Cab forward, short nose, like a Harrier jet. Stunning."

There ought to be a law against driving these three road rockets under 100 mph. Anything less, and all that engineering brilliance, all that NASA-spec hardware, all that internal-combustion splendor is simply wasted. Out here, in the arid vastness of Nevada, when the deer and the antelope don't play, we're doing our best to adhere to that unwritten rule. "Fast? This 911 is OMG-WTF-BBQ the other-two fast," says Loh of the Turbo. Kiino has a similar view.

"Don't floor the Turbo unless you've got both hands on the wheel and you're completely in a let's-go-warp-speed mode. This thing is a monster under full boil—feels like there's a jet engine strapped to the roof." Most of that speed owes to the dual-blower, 500 pony 3.8 liter six, but some is also due to the sublime seven-speed PDK twin-clutch paddle-shift transmission—which each of us agrees is about as close to perfection as the Venus de Milo or maybe even "Caddyshack."

The Big Benz, motivated by a handbuilt, naturally aspirated 8.2 liter V-8, has none of the 911's turbo lag (even with twin turbines the Porsche needs a blink to reach full power). Instead, when you plant your right foot, it simply detonates. And then you're in Idaho. "I've got it," says Loh. "The SLS actually sounds like a flaming, nitromethane-powered chainsaw cutting through a solid rocket booster." Kiino sums up his impressions of the AMG's exhaust note in one word: "eargasmic." While the Gullwing, like the Porsche, sports a dual-clutch seven speed, it's not nearly as lightning-flash quick. "Compared with the Turbo, shifts in the SLS seem to require a few extra tenths," says Kiino. "Painful milliseconds when you're in attack mode."

At the end of the article they summarize their impressions ranking the three cars.

1st Place: Audi R8 5.2: Styling worthy of an art gallery, cockpit worthy of Gucci, blistering performance and blissful civility. The German supercar you'd gamble all your quarters to own.

2nd Place: Porsche 911 Turbo: Black hole-inducing acceleration, monster grip and brakes, brilliant paddle shifter, unflappable demeanor. Incredibly, can also feel bland.

3rd Place: Mercedes-Benz SLS AMG: Huge styling drama, huge power, huge performance. But wears on you as the miles roll on, and the trickiest to handle near the limit.

Most of the information contained in this excerpt is nondimensional. The author Arthur St. Antoine uses words to paint vivid pictures that draw us into his shared experience and the experiences of his colleagues as they happily flog three very expensive German supercars in the Nevada desert. The writing style is what you would expect in a magazine devoted to car aficionados but, beyond that, it captures the spirit, fun, metaphysical and fantasy elements of what it was like to do what they did.

Certainly there were physical aspects associated with the experiences they had attached to phenomena like g force, acceleration, deceleration and kinesthesia. But the clear focus of the article was how much fun they had in driving the cars and imagining what it would be like to own one of them. All this is nondimensional and provides entrée into what goes on in the heads of car aficionados, given the opportunity to drive some of the best and most exotic cars available.

Nondimensional *seeing* is nothing like perceptual seeing. When a person who cares little or nothing about cars looks at a Porsche 911 Turbo, they see it in the same way they see a coffee cup sitting on a table. It's just another object—*out there*. For car aficionados, the experience is different. While they have the same perceptual experience as the non-aficionado did, they also *see* the car in all its nondimensional splendor. They understand and appreciate what the engineers who designed it have made available to people who value impeccably designed, high performance automobiles.

It's the same with people. Most of us have had the experience of meeting someone new and deciding we didn't like him or her. The way we *see* the person has a negative impact. That kind of reaction is based on feelings and it's nonobjective. Something about how the person looks or what they say, or how they say what they say creates a negative impression and, if it is but a chance meeting, we don't pay much attention to what happens. We forget about the encounter and move on unless someone reminds us of it. And, if that happens, we most probably remember that our impression of the individual was negative. But, if it's more than just a chance encounter like, say, being introduced to someone new in the workplace, we're embedded in a situation where, even though we've decided we don't like the individual, we will, nonetheless be forced to interact with him or her on a regular basis. What's interesting about these kinds of situations is, after a certain amount of time has passed, our initial impression of the individual changes. As we get to know him or her better, we begin to *see* them differently. This kind of *seeing* is nondimensional. What we *saw* initially turned us off. But, as we get to *see* different perspectives on the individual our opinions and reactions change. Sometimes they become even more negative but, as often as not, our initial impression changes and even reverses. This kind of seeing is nondimensional. It has little or nothing to do with information we receive through our eyes.

We have similar experiences with places and things. We *see* them in different ways and, as often as not, we don't have direct control of the events or circumstances that drive our change in perspective. It's part of being multi-celled organisms capable of reacting to dimensional as well as nondimensional experience.

The price we pay for our Western mindset with its Cartesian underpinnings is the inability to fully appreciate nondimensional experience. In thinking about all this I'm reminded of a television commercial popular in the 80's when the corporate CEO was the icon that defined the zeitgeist

The actor playing the CEO was patrician in appearance, dressed in a dark suit and getting ready to leave the office for an important meeting. His secretary handed him a portfolio and said, "Good luck." The CEO looked at her disdainfully and said, "Luck is for rabbits," as he turned on his heel and left the office. He exuded supreme confidence and his secretary looked as if she had been slapped.

The theme of that commercial resonated with the prevailing zeitgeist. Powerful and competent CEOs don't need luck.

Luck is not just for rabbits. Anyone who has been alive long enough is quickly disabused of the kind of nonsense depicted in that commercial. Luck is a nondimensional aspect of existence that everyone understands, whether or not they admit to its existence.

Few people would argue with the power and value associated with Cartesian thinking. It has brought us to where we are and helped us to become who and what we are. Where we run into problems is when we lock ourselves in to seeing the world exclusively through a Cartesian lens. For most Westerners, it's a difficult habit to break.

It's much easier for children to let their imaginations shape their experiences than it is for most adults. Children know how to and enjoy pretending. When adults pretend, they are more often than not accused of lying either to themselves or to others.

A child can imagine that a tree is a monster and experience fears every bit as real as if the tree really were a monster. If an adult does that, she's deemed soft in the head. Children can wish upon a star and be certain their wish will come true. Rational adults know that wishing upon a star is a nondimensional activity that is essentially a waste of time.

Meditation is a nondimensional activity and only a small percentage of Westerners engage in it. But, there is value in occasionally engaging in nondimensional activity even if it isn't as structured or demanding as meditation.

Find a tree and hang out with it for a while. Notice the girth of its trunk and how tall it is. Touch its bark and pay attention to its texture and color. Get close to see if it has an identifiable smell. Remind yourself that, like you, it is a multicellular organism and it is alive. It can feel and taste but, unlike you, it cannot see hear or smell anything. Unlike you, it can only *do* what each of its individual cells can do; taste and touch. It is purely chemo tactile.

Remind yourself that, for you, the tree *appears* to be *out there* but that, for the tree, it's the same as it is for each of its individual cells. There is no *out there.* Space does not exist for individual cells or for trees. Then, remind yourself that while the tree truly does appear to be *out there* when viewed from the perspective of the dimensional structure of consciousness and how it works, there is no *out there* for you either.

I now find myself in somewhat the same situation that Fuller did when he said: "The fact that I see you only as the touchable you keeps shocking me." When I look around while sitting in a chair or driving, and remind myself that there is no *out there* I feel the same kind of shock that Fuller described.

The illusion is good and persistent and, no matter how I decide to test it, it works. It appears that there really is an *out there* and that I am in it. What causes the shock is coming to understand that there is no *out there* and then trying to come to grips with that.

Looking at what is

The labels we acquire and use help us to describe and define the world we perceive. They *become* the world and, actually they *are* the world. We take them for granted and move on with our lives. But, as difficult as it may be, try to imagine what it would be like if you had no labels. If you saw a tree you would have no way of identifying it; no way to understand what it was you were seeing. If you heard a gunshot you might be troubled but you would not know what you heard. It would just be a loud, possibly startling sound. But, absent the label *gunshot* it would be absolutely meaningless. If you heard music, while you might find it pleasant you would not know what it is. It would just be a sound. The same would be true for spoken words. They would be nothing more than sounds regardless of what language the speaker was using.

You would have no way of distinguishing one experience from another. Touch, taste, smell, hearing and vision might be separate kinds of experiences but, without any labels to distinguish one from the other it would be impossible to make any meaningful discrimination between the different types of experiences you were having.

This is what it would be like if you could truly look at what is by removing or looking past the labels.

In thinking carefully about all this a question arises: if it were truly possible to remove all the labels what would be left? The answer is chaos. Looking at it, you would not be able to make any sense of it. As discussed earlier, that is what meditation allows you to do, to glimpse the chaos.

When it happens it is extremely difficult to make any sense of the experience because what you are seeing is not in time. The chaos is not in time and, when you see it you are not you. I have had this experience even when not meditating. It has happened while waiting for a traffic light to change at an intersection. But more often, it happens when I am completely relaxed sitting in the back yard and smoking a cigar. It's difficult to describe but I will try anyway.

When it happens, it is just a glimpse. And, when I try to remember how long it lasted, I have to remind myself that what I'm glimpsing is not in time. In terms of the mechanics of the experience, what I believe happens is, for some reason, the "I" stops grabbing. During meditation what probably distracts the "I" is my mantra. When the *glimpse* occurs, everything stops including thoughts and the mantra. The chaos spews a shard which, more often than not, is unrecognizable and I feel the "I" rise again which terminates the experience of separation.

Outside meditation, the experience is more strange because, when the "I" releases, I am still experiencing normal perception. I can still see what I was looking at before the "I" let go but the scene stops. Everything looks the same but like a slightly out of focus yet somehow enhanced still picture rather than what I usually see when the "I" is fully engaged.

My best guess as to why this happens is because the "I" arises in order to do but not in order to be so when it releases while my eyes are open my experience of the chaos is one dominated more by *being* than *doing*. I get a glimpse of what it is like just to be without doing anything.

I don't know why I am able to occasionally experience the "I" detach while just sitting quietly and relaxing or waiting for a traffic light to change. Perhaps it's because I have been meditating for 36 years. Whatever the reason, the experience is always novel, exciting and extraordinary. And, as much as it fascinates me, I have not been able to determine exactly what causes the experience to happen. It just sometimes happens but I have not figured out how to *make* it happen.

Different opportunities exist for those who wish to glimpse the chaos. Meditation provides a direct route but it is not the easiest route. Learning to meditate makes it possible to glimpse the chaos directly but, interestingly, even direct glimpses don't make much sense. When it happens, during meditation, while you get to see the chaos churning and spewing images and objects, for me anyway, the objects are unrecognizable. And since what I am experiencing is not in time and I am not me when it happens I cannot remember what I *saw*.

Then there is the indirect route for glimpsing the chaos. This happens to physicists when they attempt to set up experiments to examine what is, like the double slit experiment. The result does not make sense in terms of the way the world works on the macroscopic scale of existence. They release the photons or electrons, one at a time, causing them to fly toward the double slits and instead of seeing what common sense demands they must see, they see an interference pattern. And, if they try to cheat by placing a detector behind the plate with the double slits, so they can observe which slit the photon chooses before it hits the detector, then they see the pattern they thought they would see without the detector. It is a genuine WTF event for the physicists and for those who, while they may not be physicists, still understand what the physicists are seeing versus what they expected to see.

One way to interpret this result is that, while it is true that, even though what happens on the quantum level of reality is still deterministic, as observers, when we try to peek behind the veil, on the macroscopic scale, to bypass or get around the uncertainty associated with quantum objects, we are not allowed to see what those objects are doing.

In setting up the experiment, because we are manipulating the photons or electrons we become entangled with them and it is the entanglement that informs them whether or not we have placed a detector behind the plate with the double slit. On the macroscopic scale of existence, that kind of entanglement is disallowed. Einstein referred to such entanglement as spooky action at a distance. Working with two other scientists he came up with the classic EPR or Einstein, Podolsky, Rosen, thought experiment.

Quantum mechanics showed that if two subatomic particles became entangled, if you measured the spin of one of those particles, relative to an arbitrarily chosen axis, then you could know the spin of the other particle instantaneously, relative to the same axis, regardless of the distance between the two particles.

The equations of quantum mechanics showed this to be the case but, at the time, in 1935, the technology required for actually carrying out such an experiment did not exist.

The purpose of the EPR thought experiment was to show that if the equations were right, then some "hidden" variables had to be present to explain how it was possible to determine the spin of the entangled particle instantaneously regardless of how far the particles were separated. The equations were suggesting that the information regarding the spin of the measured particle was, somehow, transmitted faster than the speed of light which is strictly forbidden by the theory of special relativity. The conclusion—at the time—was that since the equations showed it would be possible to know the spin of the entangled particle, regardless of the distance separating the two particles then either the equations were wrong or there were some hidden variables associated with quantum mechanics. Neither answer was acceptable.

It is interesting, in retrospect, to contemplate how truly upset Einstein was about what the quantum mechanical mathematics was suggesting. He was upset enough to write and publish a paper with Podolsky and Rosen which, at the time, he believed refuted what the mathematics showed to be true.

The mathematics clearly showed that, if two entangled particles were separated, even by millions of light years in distance, if you changed one of those particles in any way, that change would be transferred to the other particle instantaneously. That means, according to the mathematics, the particles could exchange information about the change faster than the speed of light. It implied non-locality and that is why Einstein referred to it as spooky action at a distance.

Einstein died believing his thought experiment essentially proved that either the quantum mechanical mathematics was wrong or quantum theory was somehow incomplete. He published his paper with Podolsky and Rosen in 1935.

Almost thirty years later another physicist, John Bell, proved mathematically that the quantum mechanical mathematics were correct and, even though his algebraic non-quantum mechanical mathematical proof was highly convincing, non-believers still clung to the idea of locality and rejected the possibility of spooky action at a distance without the proof provided by an actual experiment.

Eventually we developed lasers and instrumentation robust enough to carry out such an experiment and, to the chagrin of the non-believers; the experiments showed that Einstein was wrong while the mathematics of quantum mechanics and John Bell were right.

To this day, the results of those experiments represent a bona fide WTF situation within physics. The reason these results are so baffling is that physicists, by definition, dedicate their lives to looking at what is. They choose some aspect of what they see going on and then attempt to explain what they see. One splendid example is Isaac Newton. He saw that the moon and earth tugged at one another and deduced that the cause was a force he defined as gravity and then went on to devise the mathematics necessary to describe gravity and the inverse square law. He understood that the force between the moon and the earth was the same force that causes an apple to fall from a tree. Since the time of Galileo, this is what physicists have been doing—looking at what is and trying to explain what they see. Newton's mathematics accurately described the inverse square law relationship between the earth and moon what they didn't explain was how or why the interaction between the two celestial bodies was instantaneous over so great a distance. Eventually Einstein showed, with general relativity that what *looked like* instantaneous spooky action at a distance between the earth and moon was really a result of the mass of the sun warping space–time causing the earth to fall into and follow the well caused by the mass of the sun in the fabric of space–time.

With the double slit experiments, while physicists could not explain what they were seeing, there was no indication that anything spooky was going on. They explained away the results they got by saying if you set up the experiment one way you get waves and if you set it up another way you get particles and while they couldn't and still can't explain why that's so, (there was a strong WTF factor) it was and still is considered non-spooky compared to what happens with entangled particles.

With entangled particles, all bets are off. The experimental results show clearly and unequivocally that if you make a change to one entangled particle, the other particle reflects that change instantaneously regardless of how far they are separated; spooky and WTF.

These experiments and the results physicists get when they conduct them are indirect glimpses into the chaos, into *what is* at the quantum level of reality. What makes them so important is they are scientifically reproducible and they raise serious questions regarding our understanding of how the universe works by experimentally violating locality.

If we could do the experiment using macroscopic objects nobody would believe the results. So, instead let's do a thought experiment which replicates what happens with entangled particles using macroscopic sized objects.

Imagine we each have a ping-pong ball that has two lights inside; one glows blue and the other red. One of the lights is always on in each ball and, if one ball is glowing red the other glows blue so our respective ping-pong balls are always glowing either red or blue. Each ball has a two-position switch which changes the color of the ball.

Now, imagine that these two ping-pong balls are entangled in the same way subatomic particles can become entangled so that if I flip the switch on my ball to change it from red to blue your ball instantaneously changes from blue to red. The same thing happens if you flip the switch on your ball.

Regardless of how far you and I may be from each other, the moment either one of us flips the switch on our ball, the other ball changes color instantaneously. Even if you stayed on the surface of the planet and I rode the shuttle to the space station with my ball, they would remain entangled and the color change would happen instantaneously.

At this point you may be thinking, what's the big deal? Here's the deal. The laws of physics forbid the existence of ping-pong balls that could do what I am describing in this thought experiment because, in order to pull off the color change instantaneously they would have to be able to communicate with each other faster than the speed of light. Even if you and I were only, say, 1000 miles apart and the ping-pong balls were communicating via radio waves, there would be a measurable time delay between the color change because radio waves are a form of electromagnetic radiation which, like visible light, travel at 186,000 miles per second.

The key to understanding the strangeness of such an entangled pair of ping-pong balls is there is no time delay associated with the color change when either of us flips the switch regardless of how far apart we are from each other. We could be on opposite sides of the galaxy and the color change would still be instantaneous. That is what non-locality means and that is why Einstein referred to what the equations were telling him about entangled particles as spooky action at a distance. It is also why he wrote that paper with Podolsky and Rosen known as the Einstein, Podolsky, Rosen or EPR thought experiment.

He knew that any physicist worth his or her salt would immediately be able to *see* that what the quantum mechanical equations were suggesting was impossible. Make a change to one entangled particle and the other one changes instantaneously. Einstein found the very idea of such an occurrence to be preposterous but he was wrong. He was indirectly glimpsing the chaos when he interpreted the meaning of the quantum mechanical mathematics and refused to accept what he *saw*.

Physicists look at what is and try to make sense of what they see. But, when they glimpse the chaos, albeit indirectly, with double slit experiments and entangled subatomic particles, they are unable to satisfactorily explain what they see. None of the labels they have are of any use in explaining what the experiments are showing them. Every time they conduct one of the experiments they get a glimpse of something that happens on the level of reality where chaos exists—the quantum level of reality.

Physicists are doggedly determined. They will continue to scratch the intellectual itch caused by double slit and entangled subatomic particle experiments. And, I believe, as Samuel Avery said in Transcendence of the Western mind, when they finally come to understand more deeply what it is that they are seeing, no one will be more surprised than the physicists.

WHO ARE YOU?

Who are you? It is an interesting question, for a host of reasons but, when you begin to examine it carefully it becomes, in some respects, like questions we explored earlier about light, time, space and mass. I mean this in the sense that with light, time, space and mass, we at first believe we fully understand what they are; but, as we begin to ask questions about them, we discover that our initial concepts of what these phenomena truly are do not serve us very well in terms of gaining understanding that is both deep and true. So most people simply make assumptions about light, time, space and mass and get on with living their lives. Those truly interested in understanding these phenomena deeply must wrestle with the enigmas that arise when the intellectual probing becomes focused and sharp.

And so it is with the question "who are you?" Most people when asked respond with their name. If you are alone and ask the question of yourself you might look in a mirror. The mirror reveals a reflection of what others see when they encounter you. But, as you look at your own reflection, the question, "is that really me?" arises and the answer is it depends.

What you see in the mirror reveals your physical self or, as Buckminster Fuller said the *touchable* you but it does not reveal anything about the abstract, metaphysical *thinkable you.*

The touchable you exists in the dimensions. The abstract, metaphysical thinkable you exists outside the dimensions.

The touchable, tangible you will eventually die and be put in the ground. The metaphysical/thinkable you cannot die nor can it be put in the ground. The metaphysical/thinkable you exists, for other observers, on their image screens. And so it is for you. The metaphysical/thinkable you exists on your image screen. Along with you, the metaphysical/thinkable representations of all your friends and acquaintances exist on your image screen.

The extent to which any individual makes a lasting imprint is tied to the legacy he or she leaves behind and that legacy, by definition, is created and maintained in observational consciousness.

Individuals are pattern integrities. You are a pattern integrity.

When he wrote *Synergetics*, R. Buckminster Fuller was attempting to explain his understanding of what he referred to as "scenario universe." Pattern integrity is one of the terms he invented to facilitate his explanation.

Fuller saw the universe as a constantly unfolding dynamic event. Everywhere he looked he saw dynamism. Solitary objects are *events* in the vernacular of *Synergetics*. Instead of simply existing they *happen*.

One way to bring some understanding to the meaning of the term pattern integrity is to examine how it applies to something as simple as a knot. If you take a piece of clothesline line and tie a simple overhand knot in it, you have applied a set of instructions for creating that knot:

- Hold both ends of the clothesline
- Make a loop by crossing one end over the other tracing a full circle
- Pick up the end that lies underneath and go through the opening linking a second loop with the first
- Pull on the ends of the clothesline

The procedure of tying the knot, using the instructions, creates a pattern in the clothesline that becomes visible once you've finished.

This overhand knot pattern has integrity because it wouldn't matter what kind of material you used to create it. It would be the same if you tied it into a piece of fishing line, a length of wire, a rope made of hemp or a rubber band. The knot isn't the lump you can see and touch once you've tied it. It is a weightless metaphysical pattern integrity that becomes visible in whatever material you used to tie it. Once tied it stays put which is also part of its pattern integrity. And, like the thinkable/metaphysical you the pattern integrity of the overhand knot does not exist in the dimensions; but also—like the touchable you its pattern integrity becomes visible and touchable in the dimensions once it has been tied in some kind of material. It is only then that we get to see it on the photon screen and touch it on the quantum screen. Otherwise, if we want to *see* it we must envision it on the image screen.

If, instead of picking up the end that lies underneath and going through the opening, after you made the first loop, you just pulled on the ends of the clothesline, without pulling the end through the loop, it would disappear. The pattern integrity would not be preserved.

> 505.201 "A pattern has an integrity independent of the medium by virtue of which you have received the information that it exists. Each of the chemical elements is a pattern integrity. Each individual is a pattern integrity. The pattern integrity of the human individual is evolutionary and not static."[31]

Your physical body is to your pattern integrity as the knot in the clothesline is to the pattern integrity of an overhand knot. Just as the knot isn't the lump you can see and touch in the clothesline but rather a weightless pattern integrity; your body that you can see touch and feel isn't really you. Like the knot, you are a pattern integrity but, unlike the knot your pattern integrity is evolutionary and dynamic. The way you reveal your pattern integrity is by doing.

Our physical bodies are changing constantly because we are multicellular organisms. The trillions of cells that make up our body are constantly doing their own thing. Some, like the cells in our brain and our sensory organs communicate with each other. But, they do that because that's the way we are built or *wired*. By communicate, I don't mean to imply they converse like asking one another "how you doing?" They send signals to the brain which the brain then interprets as perhaps a blue dot of space–time from a retinal cell or sound of a certain frequency from hair cells in the cochlea.

On average, all the cells in our body are replaced every seven years. So, in a physical sense, you are not the same person you were seven years ago. What stays the same during this constant flux is your pattern integrity—the metaphysical/thinkable you. Your hair gets grey, your skin begins to wrinkle, your bones become brittle but your pattern integrity, while it is evolutionary and not static remains recognizable to you and others. The mechanism which drives the evolution of your pattern integrity as well as reveals it to other observers is doing. What maintains its integrity is being. We need our bodies in order to do. We do not need our bodies in order to be.

REVISITING THE MASH-UP

As you think about all this, it is helpful to reexamine some of the basic tenets of the dimensional structure of consciousness, the different realms of consciousness and how they are related to each other and to space and time. It is important to remind yourself of how the mash-up of the dimensions and of space and time bring what we experience as reality into existence and how, without that mash-up, which is the dimensional structure of consciousness, we would not be able to experience anything.

We are so accustomed to how we experience the world that we rarely think much about what happens when we do. We've acquired thousands of labels to help us make sense of what—without those labels—would be nothing but an unidentifiable cacophony of sensory input. Absent the labels we would not be able to identify anything we saw, heard, smelled, tasted or touched. Consciousness would be undifferentiated and would simply churn and weave through our minds but, without the labels, we would not be able to make any sense of the churning and weaving. We need the labels, seeing, hearing, touching, tasting and smelling to bring context and meaning to the realms of consciousness we encounter and to help us understand what they are. But, beyond that, to understand the mash-up itself we need to think about how it is that we are able to construct the world using only labels.

We are accustomed to the mash-up and how it works. The five separate realms or domains of perceptual consciousness are all related to one another and that *relatedness* is part of the mechanism that helps us to construct the world using the labels we have. Take that *relatedness* away and the mash-up disintegrates.

If the only realm of consciousness you had access to was the tactile realm, for you, there would be no such thing as space. Existence for you would be similar to that of plants or bacteria except that you would be one click below them because they are chemo/tactile giving them access to two realms of consciousness. They are able to experience both touch and taste; but even so, for them space does not exist. They have no way to perceive space and neither would you. You would not have access to any of the labels we normally associate with space so space would not exist. Remember—there is no box. If a tree falls in the forest and there is no one there to hear it fall it does not make a sound.

Potential is what drives, enables, feeds and provides continuity for the wholeness of the mash-up that is the dimensional structure of consciousness. It provides a contextual connection between the separate realms of perceptual consciousness by informing me of what *could be* experienced in other realms of consciousness when I experience something in any single realm.

When I hear the squirrel chittering in my backyard, I think squirrel—I envision squirrel even though I do not see the squirrel. The sound of the squirrel chittering implies that there is a strong potential for seeing what it is I'm hearing. And, even though I don't see him when I hear him, I know that if I look out the window I will see him. This *is* the contextual linking mechanism between realms provided by potential.

Because the experience of the *chittering* sound falls under the category *hearing*, a relationship is established to the category *seeing*, even though when I experience the chittering I am not seeing the squirrel. The reason there is such enormous new meaning associated with whatever is heard is the underlying structural relation that exists between *hearing* and *seeing*.

When we think about how this works, it is important to become and remain aware of the fact that *hearing, seeing, smelling, tasting and touching* are not simply loose categories that float about independently in the mind. They are, rather, highly structured realms of perception that we use to create the world from sounds, sights, smells, tastes and tactile stimulation.

The key to understanding this is to become and remain aware that the important relation which enables us to create the world exists between the categories themselves and *not* between the actual sounds, sights, smells, tastes and tactile impressions experienced within them.

When I hear the squirrel chittering I do not see him. But, because I hear him, there is an immediate and strong implication that I could *potentially* see him. The chittering sound tells me nothing about what the squirrel looks like only that, if I look in the place where I hear the chittering sound, I will see a squirrel. He might be fat or skinny, healthy or mangy, big or small but, unless I look in the place where I hear the chittering sound, I have no way of knowing what the squirrel looks like.

If we now focus on this relation between the chittering sound the squirrel is making and the potential of seeing the squirrel if I choose to look in the place where I hear the sound, a question arises regarding exactly what this relation is and the answer is—space. The relation between hearing and seeing is space. I hear the chittering coming from where I would see the squirrel if I looked.

Wherever I see or hear an object is also where that object may be tasted, touched or smelled. Space, then, becomes the *potential* within which *actual* perception occurs in any realm or domain of perception and the key operational relation between realms.

All five perceptual realms or domains are coordinated in space and, because that is so, *actual* perception in any realm is always structurally related to *potential* perception in every other realm.

This is how the mash-up works allowing us to create objects from sounds, smells, tastes, sights, and tactile impressions. The cherry *is* my experience of its smell, taste; redness, roundness, softness, and the sound I hear when my teeth break its skin. Absent those perceptual experiences—there is no cherry.

This structural relation between perceptual realms and their coordination in space which establishes and maintains *potential* perception in every other realm with actual perception in any single realm is the reason objects appear to exist above and beyond consciousness. Put differently, it is what causes us to believe the cherry exists regardless of whether or not we experience it in one or more of the perceptual realms.

The structural relation of space among and between the perceptual realms is substantially more fundamental than anything we perceive within them. This is how the mash-up works.

In all of this, the only true unknowable is consciousness. Samuel Avery calls consciousness "being" and leaves it at that in the hope we will understand what he means. If you think about it carefully, for us and other observers, consciousness exists in realms or domains which have labels. The relatedness between the realms is the structure of consciousness which we recognize as space and time or space-time. It is these structures—space and time—that coordinate and organize what we see, hear touch, smell and taste on the screen. If space-time is empty the screen is blank. And, a blank screen—empty space-time—absent any actual perception is, by definition, *potential* perception.

Seeing is believing

We like to see things for ourselves. Reading about or listening to a description of Niagara Falls is a very different experience from actually going there and seeing the falls. Even watching a movie of the falls on a big screen with surround sound pales in comparison to actually being there, hearing and feeling the thunder of the water under your feet and its spray on your skin and watching the reactions of others having the same experience.

The reason vision is so compelling is, when visible photons touch our retinal cells, space–time arises. Space–time is in light.

When I finished my tour of duty in the Marine Corps in 1964, I became a student in the hospital based radiologic technology program at Lutheran General Hospital in Park Ridge Illinois. While there, I learned the fundamentals of x-ray technology. At the time, the hospital was still using lead glass fluoroscopes for examinations of the upper and lower gastrointestinal system. A fluoroscope is used to watch dynamic processes like patients swallowing barium sulfate solution.

With a lead glass fluoroscope, the x-ray tube is behind the patient. In front of the patient there is a square piece of lead glass coated with phosphors that fluoresce (produce light) when bombarded with x-rays. The radiologist is able to move the fluoroscope up and down and from side to side while stepping on a pedal to turn on the x-ray tube behind the patient. Using this set-up the radiologist could ask the patient to swallow liquid barium sulfate and watch as the barium moved down the patient's esophagus and into her stomach. He could also watch as the barium flowed from the patient's stomach into her duodenum and small bowel. X-ray technologists and student x-ray technologists assisted the radiologists as they performed these and other fluoroscopic examinations.

When using a lead-glass fluoroscope, the room had to be dark because the light output from the fluoroscope was so low that, in order to be able to see the image, your eyes had to be dark adapted. Fluoroscopic images had to be looked at using scotopc vision which means using the rod cells in the retina rather than the cone cells located in the fovea. Under normal lighting conditions, we see photopically allowing us to see more detail and to see color. The fovea is where the lenses of our eyes focus images and since it is saturated with cone cell photoreceptors, we see most clearly when there is enough light. When the light levels are low, we must rely on the rod cell photoreceptors located on the periphery of the fovea. If the light level is low enough, as in lead glass fluoroscopy, our eyes must become dark adapted first, allowing us to see using scotopic vision.

Radiologists would wear red goggles for at least an hour before doing fluoroscopy to get their eyes dark-adapted. When they went into the fluoroscopy room, they would remove their red goggles so they could see the fluoroscope. If they had to come out of the room into normal light conditions, they would put on their goggles to remain dark-adapted.

Most student technologists didn't bother with wearing the goggles. They were able to see well enough scotopically so that they could assist during the procedure, but they were not able to see the images on the fluoroscope very well. I was very interested in being able to see what the radiologist was seeing so, when assigned to assist during fluoroscopy, I always wore red goggles just like the radiologists. My fellow students thought I was nuts but I wanted to see the fluoroscope.

The experience was interesting because, while once dark-adapted, though I could see the images on the fluoroscope, it was still difficult to focus on them because the light was too dim to allow vision using the fovea.

It was all we had and we did the best we could. A year later, the hospital purchased its first image amplifier for one of the fluoroscopes. I still remember how excited everyone was after it was installed. The image amplifier sat in the same place the lead glass sat but it was much larger and had an elaborate counterbalancing mechanism to aid in moving it around. It used a large tube to electronically amplify the light of the previously dim fluoroscopic image and there was an optical chain which allowed the user to see the image on the output phosphor of the image amplifier in a small mirror.

What caused all the excitement was, the room no longer had to be dark when doing fluoroscopic examinations. I found the technology fascinating and was amazed at how much clearer the fluoroscopic images were when you could see them using photopic rather than scotopic vision.

Space-time is in light and light is visual consciousness. In the dark fluoroscopy room, with severely diminished light, space-time became diminished too. Back then, I didn't think about space-time becoming diminished because I didn't know about the photon screen or the dimensional structure of consciousness but now, I have a better understanding about why everyone was so excited when we got the image amplifier. Eventually, fluoroscopy came to be done using not only an image amplifier but also a closed circuit television system set up to send the images to a television screen that everyone could watch during procedures. The era of the darkened fluoroscopy room came to an end.

Before x-rays were discovered, in 1895, it was impossible to see anything within the body without first cutting the patient open. After Roentgen discovered them, it became possible to determine whether a bone was fractured by taking an x-ray.

The way it works is the area to be examined is placed on or in front of a cassette which has radiographic film inside. The x-ray tube is placed above the cassette and, after adjusting the machine to control the exposure level, the technologist presses a button producing x-rays which pass through the patient and produce an image on the film.

I can still remember watching as the technologist teaching me positioned a patient's wrist on a cassette in the x-ray room. After adjusting the exposure factors, she instructed the patient to hold still and pressed the buttons required to produce the x-rays. I remember hearing the sound of the rotor in the x-ray tube spin up to speed as she pulled the lever on the control handle labeled "rotor" and after it came up to speed she pushed the exposure button. In 1964 the x-ray machines we had used mechanical relays and I could hear them open and close during the exposure. She then went back into the room, repositioned the patient's wrist and made another exposure. And, while I watched intently as she performed the examination, I couldn't see anything coming from the x-ray tube when she made an exposure. I just heard the whirring of the rotor and the thunk of the relays in the x-ray control panel. I followed her to the darkroom where we processed the films and brought them out into the light room.

The detail in those images was exquisite and I remember thinking to myself that the process seemed a little bit like a magic trick of some sort because, while I understood the x-ray machine was designed to and was, indeed, producing x-rays, I couldn't see any x-rays when she made the exposures. My Cartesian imprinting was very strong.

Eventually, I overcame my doubts about whether anything was really coming out of the x-ray tube; especially after I started making my own exposures and seeing the results on film after processing.

X-rays are electromagnetic radiation just like visible light. The difference between visible light and x-rays is x-rays have a much shorter wavelength and higher frequency than visible light. They both travel at the same speed, c, but x-rays, because of their shorter wavelength and higher frequency are more energetic than visible light. They are also able to pass through materials that normally stop visible light photons dead in their tracks.

The reason radiography works is the x-rays pass through the soft tissue of the body easily; tissue that is more dense, like bone, attenuates the x-rays so, when taking a radiograph of a wrist, while it's not possible to see soft tissue structures like muscle, fat or tendons the bones show up quite well on the film.

The foundation of radiologic technology is physics and an important aspect of the training program was a course on the physics of x-rays and radiation. While studying that course, I came to appreciate the importance of physics in helping me to understand how x-rays were produced, why they produced images on film and how, adjusting factors like kilovoltage, milli amperage, and time, I could tailor the x-ray beam to whatever body part I was radiographing.

I learned to use x-ray photons to create images of the inside of patients' bodies. And, in doing that, I could see what was going on inside the patients.

My life long and deep fascination with physics began with the physics course in radiologic technology. I wanted to know where the x-rays came from and how they were produced and, in learning about that process, I began to get an inkling about the relationship between energy and matter and how one form of energy could be transformed into another with appropriate manipulation.

An x-ray tube has two primary elements, a cathode which is essentially a small tungsten filament and an anode which in modern diagnostic x-ray tubes is a metallic disk made of sintered tungsten. To produce x-rays, first the filament in the cathode is heated up by sending a current through it. The current makes it glow like the filament in a light bulb. As it heats up electrons are boiled off its surface forming a kind of cloud. Then, a large potential difference is applied between the cathode and the anode in the form of a high voltage. For most diagnostic x-ray procedures the voltage is in the range of 70 to 120 thousand volts. This difference in potential between the cathode which has a negative charge and the anode which has a positive charge accelerates the electron cloud toward the anode causing the electrons to slam into the anode very hard at about half the speed of light. When the electrons hit the anode they stop abruptly and their kinetic energy is transformed into x-rays. And, while the process works it is very inefficient. 98.8% of the energy required to produce x-rays, for each x-ray exposure is wasted as heat.

Later in my career, I taught the course in physics to students of radiologic technology and lectured on the subject to both technologists and physicians.

My exposure to and constant exploration of physics taught me that while for most people, most of the time, seeing is indeed believing, it's a mistake to believe everything you see without questioning what you see—or don't see—in an attempt to better understand it. My familiarity with physics is what makes the dimensional structure of consciousness so appealing. In laying out how and why it works, Samuel Avery never throws out the physics but, instead insists on including it as its operational foundation. In addition to questioning what you see, it is equally important to be willing to question what you believe.

As important as physics is in helping us to understand what's going on around us and in understanding the dimensional structure of consciousness, it can take us only far enough to grapple with those aspects of universe we know are scientifically connected to what the physics is telling us. If we press further, we run into enigmas which the physics cannot illuminate or explain—enigmas like:

If the universe is a box and everything including us is in that box, where is the box?

Why does light behave as if it is composed or particles or waves depending on how we set up our experiments?

How is it possible for sub-atomic particles to become entangled so that if we make a measurement of the spin of one particle we can instantaneously know the spin of the particle with which it is entangled—regardless of how far in space and time those particles are separated?

Why is it that during carefully controlled experiments with photons or subatomic particles, like the double slit experiment, how the particles behave is completely dependent on whether or not we are observing what the particles are doing?

Why is the speed of light constant? Physics tells us it is constant but it does not tell us why.

Why is it that the equations physics depends on to describe physical processes work just as well moving either forward or backward in time yet we never see physical process move backward in time; like a broken egg suddenly flying back into our hands and reassembling itself in the process?

What is it that gives physical objects mass and how are massive objects able to resist acceleration; even if they are floating free and weightless in outer space?

It is at the point where we find ourselves struggling to understand the enigmas we encounter that it begins to make sense to search for new and different kinds of questions. It is also the point at which we need to reexamine our assumptions about concepts we thought we already understood; concepts like:

- Space
- Time
- Light
- Mass
- Matter
- Consciousness
- Being

We want to see things for ourselves in order to better understand how the world works but, quite often, what we see isn't enough. For centuries we believed the earth was flat because, when we looked around, that's what we saw. It looked flat. We also believed the sun and stars moved around the earth which itself was at the center of the universe because when we looked up at the sky the sun and stars appeared to move while we appeared to be standing still relative to the sun and stars. We lacked the larger context needed to be able to understand that what we were seeing was not, in truth, what it looked like.

We have learned much about light but we are still very far from fully understanding it. We can control and manipulate it in ways that allow us to take advantage of some of its properties. Lasers and optical waveguides allow us to manipulate light with exquisite precision. We paint images with light on our television, computer and smart phone screens and we use it to communicate by creating visual and audio signals with it. We have technology which allows us to slow it down and even stop it under carefully controlled laboratory conditions.

And now, we are coming to understand that light, in addition to being what physics tells us it is—a form of electromagnetic radiation—is something else. Light is visual consciousness.

One of the questions regarding light being visual consciousness is, if that's true, what are the implications of that observation? If light is visual consciousness how does that square with the dimensional structure of consciousness and how we perceive the world? And, how does it square with why we perceive the world in the way(s) we do?

In examining how the dimensional structure of consciousness works we begin to notice that, in addition to being visual consciousness, light has a space-time structure in that it is always moving through space at a constant velocity and regardless of physical circumstances, its velocity never varies. And while it lacks mass it does have momentum and energy. In addition to being visual consciousness and having a space-time structure with momentum and energy, from yet another perspective it is a form of energy and yet another consequence of its constant and specific velocity is that light is timeless. No time passes for photons traveling through space. They exist simultaneously in five separate dimensions, three of which are space dimensions, one time dimension and one frequency or time-time dimension because, as they move through space, they also tick or vibrate at a specific frequency.

Because space-time is in light the space-time structure of light is what determines and defines the space-time structure of the entire screen *quantum screen + photon screen*. Light and the visual realm of consciousness determine the structure of the screen rather than sound or any other perceptual realm because compared to the other perceptual realms, light is a nearly perfect medium for forming the basis for the whole of perception.

Our mind uses the dimensional order in which we see things as a format to inform us of what we would experience in the other perceptual realms. When we experience something in one of the non-visual realms our mind informs us that is when and where we would *see* what we are experiencing. And, when we see something we know where and when we need to go in order to hear, touch taste or smell what it is that we see.

This is the reason we experience the world as a single simultaneous and coordinated, five dimensional event rather than five separate non-simultaneous and uncoordinated one-dimensional worlds.

This coordinated experience is facilitated via the screen with visible light being the phenomenon that enables, colors, and coordinates what we see and experience in the non-visual realms.

Absent the screen paradigm we return to the box and the question "where's the box?" once again looms large. But, with the screen asking where it is or what it is "in" is unnecessary because the screen itself provides the concept of "in-ness." And the reason this is so is that the only appropriate context for the screen is consciousness in its entirety.

It is here that we need to remind ourselves of the operational linchpin of the dimensional structure of consciousness and that is—unlike with the box paradigm—which posits that everyone has consciousness, with the dimensional structure of consciousness nobody has consciousness. Instead everybody is *in* consciousness.

This is one of those unique situations where the comfortable maxim *I'll believe it when I see it* gets reversed. In order to be able to *see* the truth and beauty of the dimensional structure of consciousness and to gain some understanding of how and why it works you must first come to believe that instead of having consciousness everybody is, instead, in consciousness. The maxim then becomes "*I'll see it when I believe it.*"

This was true with coming to understand or be able to see that the earth is not flat and that it moves around the sun. Now that we believe it, no one has any trouble being able to *see* that the earth, instead of being flat, is a sphere that moves relative to the sun and not the other way around.

Modern physics is a product of Western thought. Physicists look at what is and attempt to explain what they see. But, as physics has progressed through the special and general theories of relativity and through quantum mechanics, it has outgrown its existing metaphysical foundations. The assumptions about how the universe works based on what the physicists see no longer square with the metaphysics that once handily explained what they saw.

Before we knew about the special and general theories of relativity and about quantum mechanics, almost everything we understood fit comfortably within the framework of the metaphysics that supported the classical Newtonian view of how the universe works. Afterwards physicists and normal mortals, like us, had to come to grips with concepts and realities that fell far outside the metaphysics appropriate to understanding the universe through the lens of Newtonian mechanics.

At first we assumed that observing a subatomic particle or photon was not really much different than observing a very tiny billiard ball. But, as our abilities to manipulate and observe photons and subatomic particles became better and more refined, we came to realize that while it was comfortable to assume an equivalence between photons, subatomic particles and billiard balls, our observations and experiments were proving that assumption was, not only wrong, it was absurd.

Metaphysics attempts to answer two basic questions as broadly as possible:

What is there? And
What is it like?

The existing metaphysical foundation for explaining some of the more obscure and enigmatic aspects of what physicists see when they examine light is not broad enough to help explain why it behaves as a wave or a particle depending on how we choose to set up our experiments with it but never as a wave and a particle simultaneously. It also fails to explain why its speed is constant or why it *apparently* moves through space.

If we examine the dimensional structure of consciousness carefully we are led to the conclusion that space–time unfolds in the touch of light. We are also led to the conclusion that, for the whole of perceptual consciousness, our experience is solipsistic. Put differently, there is no *out there*.

If that's true, then, by definition, there is no space through which light could possibly move. So, while it certainly looks like space–time and the universe exist *out there* and that light moves through space we need to remind ourselves that it also looks like, the sun rises and sets. Relative to the earth, the sun neither rises nor sets. It just sits there while the earth spins on its axis and orbits around the sun.

In attempting to come to grips with and better understand what we see when we examine light, the dimensional structure of consciousness with its screen model is more instructive than the idea of the box. It allows us to consider the possibility that while it certainly *looks like* light moves through space, it doesn't move at all. When scientists look for light in space-time they run into enormous difficulties and bizarre enigmas. If, instead, they decide to look for space-time within light they may come to understand light as something more than the phenomenon it appears to be.

> Light as visual consciousness is a form of being. If we come to understand dimensions within light, and doing within dimensions, we have come to understand doing within being.[32]

When scientists look at what is they make assumptions about what it is that they see. Those assumptions are based on prior knowledge, the myth(s) they ascribe to, and how their metaphysical foundations support their observations. They also rely on tradition. Often as not, this is insufficient to explain what they are seeing.

When scientists began to examine fire and attempt to explain what they were seeing, they posited the existence of a substance they labeled phlogiston. Substances that could burn contained phlogiston and, after they had burned, the phlogiston was released by the flames leaving behind only ash. The idea made sense because the ashes left behind, after burning a substance like wood or paper were lighter than the original object before it was burned.

Then they noticed that when they burned certain metals like tin, the white powdery ash left behind actually weighed more than the piece of tin before it was burned. This was during the 18th century and the conclusion they initially reached about metals and phlogiston illustrates how even serious scientists can be led astray when they refuse to let go of a theory and rely on tradition instead of searching for the truth about what it is they are observing. Instead of questioning the existence of phlogiston, they said under certain circumstances phlogiston can have a negative weight so when it left a substance, like tin, after being burned the substance was heavier than before it was burned. Given what we know today the idea is ludicrous but, at the time, it was taken very seriously.

Antoine Lavoisier conducted a series of experiments heating a piece of tin floating in water on a block of wood under a glass jar with a magnifying glass. As the tin turned into a white powder, the level of water in the glass jar rose higher indicating that some element in the air in the jar was combining with the tin to make the powder heavier than the original piece of tin; and the idea of the existence of phlogiston along with the idea of phlogiston with either positive or negative weight was abandoned.

Scientists came to understand that when a substance burns, it combines with oxygen and even with substances like wood and paper, if the experiments were conducted carefully saving all the gaseous byproducts and ashes of the burning process and taking into account the amount of oxygen used in the combustion process, the byproducts weighed more.

This is another example of how what you believe affects what you see. Eighteenth century physicists first posited and then came to *believe* that phlogiston existed. And even though they could not see it their initial experiments led them to believe it truly existed. As they pressed further with metals, what they saw conflicted with existence of phlogiston but, instead of abandoning the idea, they held fast to the existence of phlogiston assuming that for metals it had negative weight. They forced what they saw into the phlogiston theory to support what they believed until Lavoisier finally proved experimentally that what they were seeing had nothing at all to do with the existence of phlogiston. Once they came to believe that phlogiston did not in fact exist they were finally able to see the *truth* about what they were observing when substances burned. Instead of something (phlogiston) being released as a result of burning; when any substance burned it combined with oxygen or oxidized and became heavier not lighter.

All this calls into question the confidence most of us have in the statement *seeing is believing*. It also reveals the tension between the two statements *I'll believe it when I see it* and *I'll see it when I believe it.*

Most of us ascribe to the first statement most of the time. The second statement strikes us as iffy and, absent examples of how, when, where and why it applies, we stick with the first statement.

Still another important lesson to be learned from this is, when trying to make sense of what we see, as often as not, common sense gets in the way of allowing us to see what is really going on.

Dividing lines

Humans like to put things like ideas into categories and compartments. Some of the reasons we do this are that categorization and compartmentalization:

- Make certain ideas and areas of inquiry easier to deal with
- Provide separation between ideas and areas of inquiry
- Relieve us of the necessity of considering *everything* we know when examining only a portion of what we know
- Establish *boundaries* between ideas and intellectual disciplines like physics, chemistry, biology, astronomy, metaphysics, philosophy, ontology, epistemology, religion, values, and theology

But we have come to understand that these categories, while sometimes helpful, are not only hopelessly arbitrary but, for all intents and purposes, nonexistent. And worse, to the extent we insist on clinging to them; we prevent ourselves from being able to look at what is and ultimately being able to see the *truth* in what we are observing. We get in the way of being able to switch our modes of understanding between *I'll believe it when I see it* and *I'll see it when I believe it.*

Less than 100 years ago strict dividing lines and boundaries existed between areas of scientific inquiry like physics, chemistry, biology, cosmology, and astronomy.

Now we have hybrid disciplines that ignore the boundaries by blending ideas that were once considered separate. Some examples are:

- Biochemistry
- Astrophysics

- Physical chemistry
- Molecular biology
- Chemo informatics
- Astrochemistry
- Molecular genetics
- Organometallic chemistry
- Nanotechnology
- Biophysics

As we learn more and gather more information about how the universe works, boundaries and dividing lines between scientific disciplines begin to disappear. As we cross those boundaries, we begin to see things about how the universe works that we could not see before. One reason is crossing the boundaries forces us to abandon our *belief* that they truly existed before we crossed them. We switch our mode from *I'll believe it when I see it* to *I'll see it when I believe it*.

This is an ongoing process in science and, for scientists; it is the discipline of the scientific method that forces the issue; forces them to adopt new beliefs which ultimately allow them to see things they could not see before they adopted the new beliefs.

Belief is a concept we all talk about. Everyone has an intuitive sense about what the word means but very little research has been done on the subject.[33]

Belief is a mental state and there are not many such mental states that have the broad, intense and sweeping influence over our lives that belief has. In thinking about what the word means it is helpful to examine the extent to which belief differs from knowledge.

Most people make a categorical distinction between belief and knowledge but, unless examined carefully, that distinction is highly misleading. If someone says he knows the speed of light is 186,000 miles per second and that he believes the speed of light is 186,000 miles per second it amounts to the same thing.

In general conversation, we make distinctions between belief and knowledge with regard to degrees of certainty. If I'm certain about something, like the speed of light, I'm more likely to say know it is 186,000 miles per second than I believe it is 186,000 miles per second even though the two statements amount to the same thing. If we're not sure we're more likely to say we believe something to be true rather than we know it to be true. What is truly interesting about all this is most of what we know about the world falls between the extremes of knowing something to be true, beyond any doubt, and believing that something is most probably true.

Statements like *you can take it to the bank* and *it's probably true* represent the spectrum of our convictions regarding what we know about the world. They also express gradations of belief.

Our general attitude toward the words belief and knowledge represent another categorical dividing line we use when thinking about the different ways we understand the world and our relationship to it. But, as we have already observed, dividing lines and categories have a kind of slipperiness which makes them move and ultimately disappear as we learn more about what's going on around and between us.

The dividing line between facts and values establishes a categorical distinction which many take to be sacrosanct. Science deals with facts. Religion deals with moral values and gradations between right and wrong. Most people believe the dividing line between facts and values disallows use of the scientific method to evaluate religious doctrine or moral imperatives but if we examine the *moral landscape* as defined by Sam Harris, in his book entitled *The Moral Landscape* we come to understand that belief has distinct neurological and logical properties which, when considered in the context of well being, strongly suggests that, regardless of how sacrosanct we believe the dividing line between facts and values may be, it is an illusion. Put differently, no such dividing line exists or ever did exist.

The putative dividing line between values and facts; religion and science, has a long history. It is bright, deep and writ large across the landscape of human activity. Copernicus was acutely aware of it and refused to cross it by withholding publishing of his discovery that the earth was not at the center of the universe until after his death. Galileo spent the remainder of his life under house arrest for publicly agreeing with the discovery Copernicus had made.

To the extent we continue to believe dividing lines exist between values and facts we move ever closer to the edge of an abyss every bit as threatening and ugly as the abyss of solipsism.

Science and technology are making us more powerful than we know how to be and, unless we learn to wield that power with a view toward improving the moral landscape for all living things, and a willingness to cross and ultimately abandon our belief in the existence of dividing lines between values and facts, we face a philosophical risk as ugly as any we have encountered in the past.

In his book *Living in the End Times*, Slavoj Zizek (whom the *New Republic* calls "The most dangerous philosopher in the West") paints a picture of this ugliness worth thinking about and considering.

> What generates fear today is the causal non-transparency of the threats involved: not so much the transcendence of the cases as their immanence (we don't know to what extent we are ourselves bringing about the danger). We are not impotent in the face of some natural or divine Other; we are becoming all too potent, without understanding our own power. Risks are cropping up everywhere, and we rely on the scientists to cope with them. But here lies the problem: the scientists/experts are the subjects supposed to know, but they do not know. The becoming-scientific of our societies has a doubly unexpected feature: while we increasingly rely on experts even in the most intimate domains of our experience (sexuality and religion) this universalization only transforms the field of scientific knowledge into an inconsistent and antagonistic non-All. The old Platonic division between the pluralism of opinions (*doxa*) and a single universal scientific truth is replaced by a world of conflicting "expert opinions" themselves. As Beck perspicuously notes, today's threats are not primarily external (natural),

> but are self-generated by human activities linked to scientific advances (the ecological consequences of industry, the psychic consequences of uncontrolled biogenetics, and so on), such that the sciences are simultaneously (one of) the source(s) of risk, the sole medium we have to grasp and define the risk, as well as (one of) the source(s) of coping with the threat, of finding a way out—Wagner's *Die Wunde schliest der Speer nur, der Sie schlug*" (The wound can only be healed by the spear that made it) thus acquires a new relevance.[34]

This is indeed a deep, frightening and ugly abyss and yet, it is where we find ourselves, like it or not. We got here, in part, by our insistence on clinging to dividing lines between facts and values.

Dividing lines, whether between branches of science or between facts and values exist only to the extent we believe they do. We *believe* in them so strongly that we *see* them as being real. And while the dividing lines we draw between facts and values can be useful in terms of thinking about facts and values it is our belief in their reality that moves us ever closer to the edge of the abyss. These dividing lines are no more real than money and while, like money, they can and do serve a utilitarian function, it is a mistake to believe they are real.

What we believe influences what we are able to see. It is therefore important to exert care whenever we adopt and ultimately adapt to any kind(s) of belief(s) because if what we believe turns out not to be true, what we see can be and often is misleading. Even though people, for centuries, believed it to be true, the earth is not flat, stationary in the heavens, or located at the center of the universe. Its shape is spherical; it rotates on its axis and orbits around an unremarkable star and its existence, in terms of our understanding of the size and the scope of the cosmos, amounts to nothing more significant than a speck of dust.

Magic

Westerners usually associate the word magic with what illusionists like David Copperfield do. He performs spectacular illusions like making the statue of Liberty disappear before a live audience, walking through the Great Wall of China, and flying. And, while that kind of magic both confounds and delights us, we are not *really* fooled. We know—going in—that he is a magician or illusionist and, if we expend enough time and effort to discover how he performs his illusions, we come to understand how he is able to make it appear that he is able to do what it *looks like* he is doing. Once the secret is revealed, the magic disappears.

What David Copperfield and other magicians do, as a form of entertainment, conforms to the Wikipedia definition of Magic (paranormal):

> Magic is the claimed art of altering things either by supernatural means or through knowledge of occult natural laws unknown to science.

People who live in Western Civilizations understand that it is not possible to walk through the Great Wall of China, make the statue of Liberty disappear or to fly. So when they see someone do any of these things, even if the performer is very talented and clever, they assume it is some sort of trick.

In thinking about this I am reminded of the famous quote by Arthur C. Clark regarding magic: "Any sufficiently advanced technology is indistinguishable from magic".

To someone living in the 16th century, technologies like radio, television, personal computers and smart phones would, essentially, appear to be magic.

But there are different kinds of magic that, unlike what entertainers do, are not quite so easy to dismiss or figure out. At bottom, magic is tied strongly to belief and, as we have seen, belief has a strong influence on what we are able to perceive whether it is through vision, hearing, smell, taste or touch.

Voodoo, hexes, curses and the placebo effect are also forms of magic tied directly and inexorably to belief, so much so that they can and do have a direct influence on life and death. One of my favorite examples is a story related in *Space Time & Medicine* by Larry Dossey, M.D.

While he was an intern, he worked with a colleague, also an intern, who was assigned to treat an elderly man who had been admitted to the internal medicine service of the hospital. His admitting diagnosis was cancer because the man had lost fifty pounds in six months before being admitted to the hospital.

His colleague, Jim, ran the usual battery of diagnostic examinations, all of which showed negative results. Jim continued with the workup of the patient even though all the tests were negative, hoping to find out what was going on and convinced that some kind of real disease process was involved in the patient's obvious and continuing deterioration. As the workup continued, the patient became worse, becoming extremely weak and almost bedridden.

For Jim this was a vexing dilemma. A patient in his care was dying and he could not explain why. Jim told the old man, "You're dying and I don't know why!" The old man said, "That's all right, Doctor. I know I'm dying. And I know why, too. Doctor, I've been hexed."

When Jim asked the man how he knew he had been hexed, he told him that three months ago, an enemy of his had hired a local shaman to put a hex on him. The shaman got the old man's wife to clip a lock of his hair and give it to her. Once she took possession of the lock of hair, she began to work a spell. When she was ready, she told the old man and his enemy that he had been hexed and that he would die as a result of the hex.

It seemed odd to Jim and Larry that the old man never resisted this pronouncement. It also seemed clear that the old man never considered the possibility that the hex might not work. The moment he found out about it, it was as if he was already dead. He lost his will to live. He stopped eating and began to inexorably lose weight. When he came to the hospital, he came there to die.

When Jim told Larry about the hex, Larry's response was one of despair. He felt sympathy for the old man but could not think of anything to do about the situation. He felt the old man was right, that he truly *was* going to die.

Jim's reaction was different. He decided that he and Larry had to cure this old man who had been hexed, and they concocted a plan to do just that. Jim decided the only path open to them was for Jim to become a shaman and do battle with the shaman who had hexed the old man. Jim decided to pit his medicine against the hex the shaman had cast against the old man; and the stakes of the battle were high indeed, the life or death of the gray-haired old man.

On a Saturday night when there was a lull in hospital activity, Jim decided to hold a *ceremony* at midnight. He went into the patient's room and assisted him into a wheelchair, because by then the old man was barely able to walk.

Jim had instructed Larry to go to the examination room, get a methanamine tablet, put it in an ashtray and ignite it. Methanamine was a low potency antibiotic used to treat urinary tract infections, readily available in the hospital, and Jim and Larry knew it was flammable.

Jim wheeled the old man into the examination room and Larry locked the door. The three of them were alone in the darkness with only the flame from the burning methanamine tablet providing illumination. For a long time, Jim sat near the flame and said nothing. Then he arose from the chair totally at ease and clearly in command of the situation. The old man was transfixed.

Jim had become a shaman and he approached his task with dead seriousness because he knew the old man's life literally depended on what he was about to do. Jim took a pair of stainless steel surgical scissors from his lab coat pocket and light from the burning tablet glinted off the shiny blades. The old man watched intently, following the slow and deliberate moves Jim was making. He ceremoniously raised the scissors as he walked toward the wheelchair. Then he grasped a lock of the old man's hair and began to slowly cut.

It looked, to Larry, as if the old man had stopped breathing. Then Jim took the lock of hair to the desktop, looked at his patient and said in a calm, deep voice, "As the fire burns your hair, the hex in your body is destroyed." Then he dropped the lock of hair onto the burning methanamine tablet. He told the old man that if he revealed this ceremony to anyone, the hex would return immediately, even stronger than before.

Jim wheeled the old man back to his room and helped him get back into bed. The old man said nothing.

The effect of the de-hexing ceremony was almost immediate. Even though the hospital's breakfast menu was notoriously bad, the old man ordered a triple serving. After that he continued to order double servings for every meal regardless of what was on the menu. He began to gain weight at a spectacular pace.

The old man, as Jim had warned, never mentioned the de-hexing ceremony, even to Jim. He was cheerful and happy from that time forward. The old man was convinced that Jim—the shaman—had rescued him. Jim kept him in the hospital for several days to make sure the de-hexing had truly worked. When he was convinced his patient had been cured, he discharged him. The old man exited the hospital a well man and left behind a thick chart with normal test results. The examination room Jim and Larry had used for their ceremony reeked of burning hair for several days.

This is a different kind of magic than the kind that magicians and illusionists perform for entertainment. In this particular instance, the magic was strong enough to have a direct impact on the patient's life.

From the way the story was related by Dossey it seems clear that, without his colleague Jim's intervention, their patient would surely have died. It is also clear that the de-hexing ceremony literally saved the patient's life. The original hex had a nocebo effect on the patient and threatened his life. The de-hexing ceremony exerted a placebo effect which counteracted the original hex and saved his life. In both situations, it was the patient's belief in the effectiveness of the hex and the de-hexing ceremony that made the magic work. He saw what he believed and, in both instances, that belief determined the outcome of his situation.

For most people, puppies and kittens engender yet another kind of magic. The magic arises from the delight that interacting with them creates. There seems to be a mechanism associated with our neurochemistry that makes us predisposed to experiencing delight when interacting with puppies and kittens. For many, the effect is so strong that just looking at pictures or images of puppies and kittens provokes feelings of delight as strong as actually interacting with them.

And then there is the kind of magic associated with falling in love. The magic associated with romantic love is so strong that individuals who experience it are willing to do things that they would never even consider outside the thrall of intense romantic love; things like losing entire fortunes and committing murder.

Those who have not experienced intense romantic love cannot understand how or why being in love can make one's focus and attention on his or her beloved so intense that nothing else seems to matter. Those who have, while they too may not be able to explain or understand why what they feel is so intense, do understand what love feels like. They also understand what I mean when I say that falling in love is a genuine kind of magic.

And finally, there is the term used to explain phenomena which resist explanation using any imaginable approach. These kinds of phenomena are often explained as working on the FM principle—FM standing for F#@&ing Magic.

Magicians like David Copperfield are skilled at performing grand illusions like flying and close up magic like making coins or other small objects disappear simply by manipulating them. They are able to make us believe that what they do is magic by using techniques like misdirection and forcing. By relying on how our brains are *wired* in terms of the mechanics of perceptual consciousness, magicians are able to misdirect our attention. When he places a coin in the palm of his left hand and closes his fingers over it, even though he palmed it in his right hand, he uses it to point at his left hand—where he wants us to look and, while waving his right hand magically over his left hand while opening his fingers, we're amazed that the coin isn't there.

He uses his skill at misdirection to make us see what he wants us to see and our brains, because, of the way we're wired, cooperate so we end up being surprised and amazed when we don't see what we expected to see or when we see something unexpected.

Neuroscientists have become so intrigued by how magic works that they have created a new branch of cognitive research called neuromagic. Neuroscientists know about how the brain is wired and those exploring neuromagic have gotten skilled magicians like Randi and Penn of Penn and Teller to show them how classic magic tricks are done, like the cups and balls routine. The idea being that once they understand the mechanics of the trick, they can gain a better understanding of what's going on in your head when you see such tricks performed.

What I personally find intriguing about magic, especially classic routines like the cups and balls is, even though I not only know how the trick is done but have performed it myself in front of mystified onlookers, when I see a consummate professional do the trick I'm still delighted and amazed; especially when they add an unexpected element like loading a large object under one of the cups like a baseball or an orange.

Our experiences and memories determine and shape what we expect to see, hear, and feel. So when I see someone put a ball under a cup I expect that when he lifts the cup the ball will still be there. When it's not there, that's when the magic happens because the magician misdirected me.

As I think about this and how much I enjoy watching magicians perform I remind myself that the way my brain and everyone else's brain is able to map photons touching retinal cells into objects that appear to be *out there* is the greatest magic show anyone has seen or ever will see, hear, smell, taste or feel. The dimensional structure of consciousness is the greatest magic show on earth because, as Samuel Avery observed—"the universe unfolds in the touch of light."

NEURONS FIRING

When a neuron is stimulated it fires causing a signal to be sent along neuronal and axonal pathways. Within the body of each neuron, neural signals travel electrically. Each neuron is composed of a neural body or soma, an elongated projection or axon and branch like structures called dendrites which exist between neurons and allow neurons to communicate with each other and form groups of neuronal firing patterns. And, while neural signals travel electrically within the neuronal body and axons, they are transmitted between neurons by specific chemicals called neurotransmitters, one example being acetylcholine.

Neurons firing are responsible for a vast array of processes required to keep us alive and thriving. They send signals to our muscles and glands and they mediate all five of our perceptual senses.

When someone taps you on the shoulder, you feel it because neurons in the muscle and skin where you were tapped fire sending a signal to your brain. And it is the same with the other four senses. When photons touch the retinal cells in your eyes, since those cells are a type of neuron, they fire sending a signal to your brain and then, another kind of magic happens. Neurons within your brain begin to fire which allows you to see and causes space–time to arise. The activity initiated by the retinal cells firing and the neurons in your brain firing, as a result, provide the operational framework which underlies and is required for you, and every other observer with vision, to become aware of and take advantage of the dimensional structure of consciousness.

When certain specific molecules come in contact with the taste buds on your tongue and in your throat or the olfactory receptors in your nose, neurons fire sending signals to your brain causing another cascade of neuronal firing within your brain which you ultimately perceive as taste and smell. And, when mechanical vibrations of air molecules within the appropriate range, strike your eardrums those vibrations are transmitted to hair fibers in your inner ear which fire sending signals to your brain which, again, results in another cascade of neuronal firing within your brain which you ultimately interpret and perceive as sound.

Beyond the neuronal firing associated with perception and the dimensional structure of consciousness, lies a constant symphony of neuronal firing within your autonomic nervous system which keeps you alive by maintaining homeostasis. Your brain is constantly monitoring what is happening in your body. If you become frightened, another result of neurons firing, your brain sends signals to your adrenal glands along the appropriate neuronal path causing adrenalin to be pumped into your bloodstream.

If your blood pressure rises or falls outside the parameters compatible with life a cascade of neuronal firing takes place in an attempt to restore it to levels appropriate for homeostasis. When you eat, your brain senses the need for insulin because neurons fire informing it that your blood sugar levels have risen. Then more neurons fire within your brain sending signals to your pancreas to release the required dose of insulin. Fatigue, sleep, thirst, and hunger are all mediated by neurons constantly firing in your brain and autonomic nervous system and, for the most part, we remain unaware of all that activity until and unless something goes wrong.

If we become cold enough to freeze, neurons begin to fire sending signals to our brain and autonomic nervous system. Then more neurons fire which causes our blood vessels to constrict forcing the blood volume from our limbs into our trunk to protect our vital organs as long as possible.

We know all this because physicians and scientists have studied these phenomena in detail and mapped the neuronal pathways associated with our perceptual and autonomic nervous systems.

Using imaging techniques like functional MRI or fMRI, physicians and scientists have identified areas within the brain associated with phenomena like vision, hearing, smell, taste, touch, fear, music appreciation, mathematical ability, pleasure, pain and even whether or not we are telling the truth.

Everything we experience, whether we are aware of it or not, is the result of neurons firing.

When we examine neurons individually we discover that, in terms of the signals and potentials they produce when stimulated, the results are relatively uniform. There are over 100 billion neurons in the average human brain. Each neuron has about 7000 synapses associated with it so the average adult has somewhere between 100 to 500 trillion synapses. The question then becomes, how is it possible that neurons and synapses which fire in one area of our brains result in vision and in another area, sound? And the same question stands for everything else we experience like taste, smell, touch, fear, love, hate, disgust, delight and desire. It also applies to the activity associated with our autonomic nervous systems. How is it that the neuronal and synaptic firing that goes on results in maintenance of homeostasis? The answer is we do not know.

We have made some interesting guesses about consciousness. Most believe everyone has consciousness. Some of us, like Samuel Avery and I, believe that instead of having consciousness we are instead all *in* consciousness. The answers to the questions regarding consciousness exist on that continuum between knowing and believing but, nobody really *knows* in the same sense that we *know* the speed of light is 186,000 miles per second in a vacuum and $E=mc^2$.

The questions regarding how neuronal firing results in all the phenomena associated with perception, consciousness and maintenance of homeostasis are interesting because they are squishy. Answers lie on the continuum between knowledge and belief and those with a scientific bent prefer knowledge to belief. However, the information available to us regarding exactly what is truly going on in all this neuronal and synaptic firing lies much closer to the belief side of the continuum than the knowledge side; particularly when the questions concern consciousness.

As we have seen, the subject of consciousness is very squishy because dealing with it requires the inclusion of phenomena like qualia. The moment that happens scientists become uncomfortable and apologetic. Exploration of phenomena like qualia become rapidly entangled with philosophical issues and, to a large extent that is what fuels the discomfort scientists feel and exhibit when they talk about subjects like consciousness and qualia. But, as we have also seen, even scientists who pursue the hardest of the hard sciences—physics—are forced to deal with issues that become as squishy as and spookier than consciousness and qualia.

The typical reductionist approach to a problem like trying to determine how neurons firing can result in perception, consciousness and maintenance of homeostasis is to break the problem down into its smallest parts, examine those parts and see how they all fit together. And while that approach is both useful and powerful, for certain problems it just does not work.

On the quantum level reductionism loses its power and allure. When we send photons or even subatomic particles like electrons, which have mass, through a plate with two slits cut into it our most careful observations and measurements yield results which defy common sense. We are forced to conclude that the particles take *every* path possible on their way to and through the slits allowing them to interfere with themselves, on the other side of the plate with the slits, resulting in an interference pattern and if we observe which slit the particles go through, the interference pattern disappears. And, if scientists purposely entangle two subatomic particles, they have observed that regardless of how far those particles are separated any change made to one particle has an instantaneous effect on the one with which it was entangled; even though special relativity forbids such a result.

Physicists, especially those fond of reductionism, abhor uncertainty because it is squishy. But, thanks to Heisenberg, they must find ways to deal with it.

In his book *I Am A Strange Loop*, the well-known and celebrated polymath Douglas Hofstadter uses an approach to explain what is going on in the brain during all its neuronal firing that Einstein often used when faced with squishy questions—the thought experiment.

His thought experiment is based on a metaphor he created for thinking about how the brain works and explaining the myriad levels of causality in our brains minds and possibly even our souls. The name he used for this metaphor is The Careenium.

To begin to work with and gain some appreciation for the careenium Hofstadter asks us to imagine a large and elaborate surface something like a pool table. Then he asks us to imagine that on it there are billions of tiny silver marbles he calls *sims* which is an acronym for small interacting marbles. Then to make things interesting he posits that the surface of the pool table is perfectly flat and frictionless and that the sims are frictionless too. Now, because the sims and the environment they are in is frictionless and perfectly flat they constantly bash into each other and the walls of the careenium never stopping. At this point any good physicist would tell you that Hofstadter's description of the frictionless sims on the frictionless surface of our metaphorical pool table is the equivalent of an ideal two dimensional gas; so far so good. Then he adds another level of complexity by asking us to imagine that the sims are magnetic, adding an extra m to their name, changing it from sims to simms—a new acronym for small interacting magnetic marbles.

Now, when the simms bash into each other, if they are moving slowly enough they will stick together forming clusters. He calls such clusters simmballs. If you are not familiar with Douglas Hofstadter you are probably beginning to understand that, in addition to being a brilliant polymath he is clever, ironic and funny.

Simmballs are composed of large numbers of simms in some cases hundreds of thousands and in others millions or hundreds of millions. On its periphery each simmball frequently loses some of its simms while simultaneously gaining some. The careenium has gained some more complexity. Now it is populated by two different kinds of objects: the tiny almost weightless simms constantly bashing into each other and the sides of the careenium and the giant ponderous simmballs which move so slowly they appear to be immobile.

The operational dynamics associated with the careenium consist of simms zooming around, crashing and bashing into each other, the sides of the careenium and into the simmballs. There are lots of collisions going on constantly.

If you've gotten this far in reading about the careenium you've also most probably picked up on Hofstadter's corny pun on the word symbol. He did that for a reason—to add even more complexity to the careenium. He then posits that the vertical walls which make up the careenium are extremely sensitive so they react to outside events like someone touching one of them or even gently breathing on one of them by flexing inward. Each time one of the walls flexes some traces of the nature of whatever event caused the flexure is retained by the section of the wall that flexed and, since the walls are constantly in contact with the simms the simms are affected by the sections of the walls that experience any kind of external event. When they bounce internally off the section of the wall that flexed this will be registered, indirectly, in the very slow motion of the nearest simmball allowing it to *internalize* the event.

Now if we assume that a given simmball always reacts in some predictable or standardized fashion to events like, say, a tap, a breeze, a vibration, a sharp blow or whatever we can also imagine that the configurations of the simballs reflect the history of whatever kinds of events impinged on the walls of the careenium from the outer world.

If it were possible for someone who knew how to read and interpret their configurations, to look at the simballs from above the careenium, by definition, their specific configurations would be symbolic of the events that encoded them. Simmball = symbol, get it?

Certainly, there is no such thing as a careenium but, this is a thought experiment so that doesn't really matter. The careenium is intended as a useful metaphor for understanding how the brain works and while it is certainly mind-boggling so is the fact that our brains also contain tiny events, in the form of individual neurons firing and larger events in the form of patterns associated with groups of neurons firing.

Tests and diagnostic procedures like fMRI provide strong evidence that patterns of neurons firing encode and symbolize events happening outside our crania and, as outlandish as that may seem, we are led to the inescapable conclusion that all this happens as a result of evolutionary pressures ultimately geared toward helping us to survive.

The salient points regarding the careenium and how it works attach to the metaphor of simms with individual neurons firing and simmballs with patterns of neurons firing. With the careenium we know that the simms, on their own, do not encode anything nor do they play any kind of symbolic role while, the macroscopic simmballs do encode events and are, therefore, symbolic.

A hardnosed physicist with a strong bent toward reductionism might look at the careenium and conclude that the only components that matter are the simms with the simmballs being mere epiphenomena. Put another way, while the simmballs are certainly there, on the careenium, they're unimportant to truly understanding the system because, after all, they are composed of simms. Therefore, ipso facto everything that happens within the careenium is, by definition, explainable in terms of the simms alone. From a strictly reductionist perspective, certainly this is true.

If I look at the rectangular block of dull gray titanium that sits on my desk, like the simballs, it too is undoubtedly there. If we examine it using a purely reductionist approach we needn't talk about or concern ourselves with its size, how light it is compared to steel, its tensile strength, its hardness or its ability to resist corrosion. These are, after all, simply epiphenomena that we can dispense with. We can best understand the block of titanium by focusing on what it's made of—protons, neutrons and electrons. From a purely reductionist perspective, epiphenomena are nothing more than convenient explanatory tools used to summarize larger numbers of lower level phenomena. They are therefore, by definition, unessential to any explanation.

The problem with this approach is how far and fast it escalates the complexity when we choose to ignore the normally macroscopic ways in which we look at things like blocks of titanium, rocks, clouds or trees. Refusal to use language involving epiphenomena forces us to look only at the sub-atomic particles which make up what we are examining. The question then becomes, how helpful is that in terms of gaining any meaningful understanding of what it is we are examining?

Looking only at sub-atomic particles all the sharp borders normally associated with our usual macroscopic experience of the world disappear. We wouldn't be able to define the volume of the block of titanium and say, for instance, this is where the right side edge exists because sub-atomic particles are identical. They have no respect for epiphenomena like edges, color, tensile strength, or surfaces. The same would be true for water. Even if we choose to move up a notch from sub-atomic particles to atoms and molecules, there is nothing inherent in hydrogen atoms or oxygen atoms alone to suggest that when two atoms of the gas hydrogen combine with one atom of the gas oxygen that, on a macroscopic level we should expect a fluid that is wet. A single naked water molecule would not be wet.

All this exposes the problems inherent in using the reductionist approach to examining epiphenomena with which we are familiar and which have a serious impact on us and the way(s) we live our lives. For someone who doesn't know how to swim, who suddenly finds himself in deep water, knowing that it is composed of hydrogen atoms bound to oxygen atoms will not be very helpful.

Since, by definition, there is a comprehensible logic associated with the kinds of events we encounter as epiphenomena, we humans gladly jump to that level. We do talk about the size, shape, tensile strength and weight of the titanium block and how it might be better than steel if what we want to build must be lighter, stronger and have greater tensile strength than we could get using steel. We see the world around us in terms of the epiphenomena we encounter and talk about things like cars, sour dough bread, hardwood floors, pimples and people with weird ideas like Stephen J. Hage. The most obvious reason we do this is that we, ourselves, *are* epiphenomena.

Let's return to the careenium. On it we have the simms zipping around, bashing into themselves and everything else on its surface and its walls. And, while the simmballs are also on the surface of the careenium, they are just these big stationary objects which the simms bash into and bounce off of. In that sense they are much like the walls.

But if we choose to look at the careenium by shifting our perception of it in two important ways we arrive at a picture quite different from the one we started with.

The first perceptual shift involves imagining that we're watching what's going on as if it were a time-lapse movie. When we do that, imperceptibly slow motion gets speeded up allowing us to see it while fast motions become so fast we can't see them at all. They become blurred like the spinning blades of an airplane propeller.

If we then change our spatial perspective by backing away or zooming out the simms disappear completely and our attention becomes focused only on the simmballs.

This forces a radical change in the dynamics on the careenium. We no longer see tiny simms zipping around crashing and bashing into each other, and the large stationary simmballs. We also immediately notice that the simmballs, far from being stationary, appear to be very lively, moving around the surface of the careenium and interacting with each other as if they were the only objects there.

And, while we know that, on a deep level, what we see happening is because tiny simms are zipping about we're not able to see them anymore. In our speeded up, zoomed out view of the careenium the simms have become just a stationary gray background against which we see the simmballs moving around and interacting with each other.

If it were possible to shift our perceptual gestalt in the same ways we did with the careenium while looking at a glass of water, allowing us to see it at the micro level, we would immediately realize that it's nowhere near as quiescent as it appears on the macro level. We would see the individual water molecules constantly crashing and bashing into each other and the sides of the glass. If we then poured some cream into the water we would see Brownian motion because cream in water is a colloidal suspension and we would notice immediately that the larger molecules of cream were being randomly pushed around by constant collisions with the water molecules. Here, the larger molecules of cream are the simmballs and the water molecules are the simms.

For a long time Brownian motion was a mystery. It was named for the English botanist Robert Brown who, in 1827, while examining pollen particles suspended in water, under a microscope, noticed that they moved around wildly in a zigzag pattern. In the 19th century we didn't know about atoms or molecules. Even as late as the early 20th century the existence of molecules was hypothetical. In 1905 Albert Einstein explained Brownian motion using the theory of molecules and his explanation, because it was consistent with experimental data, was so far-reaching it became one of the most important confirmations that molecules truly exist.

We've now reached the point where we can begin to ask some questions about the careenium linked to how we choose to look at it and, in so doing, attempt to determine which view of it is the truth. The answer is ultimately linked to the question Roger Sperry asked about it: "*Who shoves whom around in the population of causal forces that occupy the careenium?*"

The reductionist view sees the simms, which have no intrinsic meaning, as the primary entities madly zipping around and, as they crash and bash into each other, they push the bigger and heavier simmballs around. And, since the simmballs are themselves composed of simms they are not recognized as separate entities. Talking about or focusing on what they do is simply another way of talking about or focusing on what the simms do. For the reductionist it makes no sense to even talk about simmballs, symbols or thoughts being encoded because all there is on the careenium, is tiny shiny magnetic spheres that careen about madly in a way that's essentially pointless.

But, when we speed up what's going on and zoom out the simms disappear morphing into a featureless gray fluid and our attention becomes focused on the simmballs because, in this view, they look like they're interacting with each other in interesting ways. Groups of simmballs appear to have an impact on other simmballs suggesting a kind of "logic" unconnected with the featureless gray fluid that's constantly churning around them other than providing energy. The logic we perceive in the interactions and motion of the simmballs is associated with the concepts they symbolize, not with the simms.

When we look at the careenium using our zoomed out, speeded up, macroscopic perspective, what we see is thoughts and ideas giving rise to other thoughts and ideas. We can see a given symbolic event reminding the careenium of another symbolic event which took place elsewhere. We can also see intricate patterns of simmballs coming together to form even larger patterns that make up analogies. What we're doing here is eavesdropping on the logic of a thinking mind occurring in the patterned dance of the simmballs. And, from this perspective, it is the simmballs that shove each other around at the isolated symbolic level they possess.

Certainly there is no question regarding whether or not the simms are there—they are. But, all they're doing is serving the simmballs' dance and allowing it to happen. The micro details of their individual bashings have no identifiable relevance to the processes associated with cognition. The situation is similar to the micro details of air molecules crashing and bashing into a child's pinwheel toy causing it to spin. For the pinwheel, any old crashing and bashing will make it spin because of the aerodynamics associated with its design. And so it is with the careenium. The *thought-mill* will churn regardless, driven by the symbolic nature of its simmballs.

The idea of the physical existence of a careenium is certainly improbable but if I think about the human brain and ask myself what goes on in it which allows our thinking's logic to take place I end up asking the question Hofstadter asks, "What else is going on inside every human cranium but some story like this?"

The question regarding who's shoving whom around remains. The answer depends on the level we choose to focus on.

We need a clear sense of the relationships between different levels of description when attempting to bring some understanding to descriptions of entities that think. And it is here where the careenium is helpful. On one level all we have is simms crashing and bashing about in ways that demonstrate no logic and make no sense and the simmballs are nothing but stationary large objects composed of simms—not very interesting. But, from a different perspective, we find a level where the meanings attached to various simmballs *what they have encoded* can quite legitimately be understood to be shoving other simballs around.

All this may seem wild and highly improbable but what makes it interesting and worth considering is how consistent it is with fundamental causality and the laws of physics.

WEALTH AND MONEY

Most people understand what the word "wealth" means. Dictionary.com offers this definition:

Wealth

noun:

1. a great quantity or store of money, valuable possessions, property, or other riches: the wealth of a city.
2. an abundance or profusion of anything; plentiful amount: a wealth of imagery
3. Economics .
   a) all things that have a monetary or exchange value.
   b) anything that has utility and is capable of being appropriated or exchanged.
4. rich or valuable contents or produce: the wealth of the soil.
5. the state of being rich; prosperity; affluence; persons of wealth and standing
6. Obsolete: happiness.

Few would argue with this definition because most dictionaries yield similar interpretations of the meaning of the word "wealth." We associate wealth with the possession of money and the things we can buy with it and, for most of us, most of the time, that suffices. But, how we see something depends, in large part, on how we choose to look at it, as we saw earlier with the careenium.

R. Buckminster Fuller defined wealth as energy compounded by knowledge. But he went further than that. He said if wealth is understood as energy compounded by knowledge then it is not only inexhaustible but it must always and only increase. The reasons he provided are linked to the components.

Physics tells us that the total amount of energy available in the universe is constant, via Einstein's famous equation $E = mc^2$. Energy is, therefore by definition, inexhaustible.

With knowledge, Fuller pointed out that it is not possible to learn less. As long as we remain alive and viable we always learn *more* about whatever we might be exploring or learning how to do. If you've never played tennis and decide you'd like to learn, during your first few trips to the court, your performance will be abysmal. You will regularly miss hitting the ball and fail to keep it within bounds when you do hit it. But, if you stay with it, and keep going to the court to practice, eventually, your skills will improve because you will be learning *more* about what it takes to be successful at the game. You will learn how to anticipate where the ball will land when it is served or returned and how to place it when you hit it with your racket. If you continued going to the court and practicing, it would be impossible for you to learn less about the game of tennis.

So the second component of wealth, as defined by Fuller, always and only increases which is why wealth, defined as energy compounded by knowledge, can only increase.

Perception defines reality. Which definition we choose shapes our perception of the word and the extent to which we believe it to be true.

Fuller's definition gives us reason to rejoice.

The more conventional definition creates a different understanding. Wealth becomes tied to money and is something to be acquired, possessed, and tended to carefully, lest it be squandered and lost.

The problem with money is, as pointed out earlier—it isn't real; even though most of us, most of the time, behave as if it is real. We go to work, put in our time and, at the end of each pay period, receive a check in the form of a paper document we can take to the bank and either deposit or exchange for cash. And, for most of us, we don't even receive a check. Instead we get a document which tells us that the amount we earned has been electronically deposited in our checking account. In reality, nothing physical is deposited into our accounts. Our employer's bank sends an electronic message to our bank telling it the amount of our check. The message is nothing but 1s and 0s, represented by the magnetic polarization of microscopic domains on the bank's hard drive which, most likely, isn't even located at the bank where we do our banking.

If you use a debit card when you make a purchase, more 1s and 0s are transferred between your bank and the merchant's bank. Nothing *physical* ever is exchanged.

One reason this is so is, it's easier to do the accounting and keep track of transactions by shuttling 1s and 0s around on the surfaces of hard disks than it is to do the same with physical cash. That's the reason we've moved from cash to credit and debit cards and the various forms of electronic funds transfer (EFTS) available, worldwide. Electrons move more swiftly, at nearly the speed of light, and can be handled much easier than physical cash.

When I think about all this and how truly ephemeral money has become, in society at large, I like to engage in a thought experiment about the situation. I imagine that some kind of incident occurs which instantly and irretrievably wipes clean every hard drive which stores financial information of any kind. Since this is just a thought experiment it doesn't matter what kind of incident it was. Then I ask myself, since all the financial information regarding how much money everyone has is no longer available how would people react? The situation raises many interesting questions about what the world would be like; for instance:

> Could anyone be considered wealthy absent the electronic information which previously existed?

Certainly some people would have paper records of the accounts in which they had money prior to the incident which wiped all the hard drives clean but, even so, there would be no way to verify the validity of those documents. Others might have some information stored on thumb drives or optical media of some sort but, once again, it would be very difficult if not impossible to verify the validity of that information. Our computers would still work the same way they did before the incident occurred so all the same hackers and phishers and spoofers who existed before the incident would be free to *create* electronic information regarding how much money they had before all the information was wiped clean.

Would there be any money in the world?

There would still be cash in the form of paper money and coins but, for most people, that wouldn't be of much help. And, even for the banks and other institutions and individuals who had cash, there would be no way for them to verify that it truly belonged to them. Banks and financial institutions deal exclusively with other people's money so any cash they might have, on hand, would, by definition, belong to other people not them.

Governments and treasuries can print money but, without access to any verifiable information regarding who had how much money prior to the incident, there would be no way to allocate the money they printed. Until and unless someone found a way to restore the information that was lost, by definition, nobody would have any money other than those who happened to have some cash when all the hard drives were wiped clean. And while most people would be without money they would also be without debt.

The situation would be interesting, difficult and chaotic. Major businesses like, General Motors, IBM and General Electric would have no money. But what they and every other business would have is the physical and human resources they had prior to the incident.

Some businesses like Microsoft and IBM which rely almost exclusively on knowledge workers to produce their products would be well suited to continue doing what they did before all the financial information was wiped clean. Their programmers and systems engineers would still be able to write code and produce new software.

Businesses engaged in manufacturing like General Motors and General Electric wouldn't have it so easy. They need stuff in order to be able to make their products but, without access to money they wouldn't be able to buy the materials they needed. And no business would be able to pay their employees.

People across the globe would still be able to do what they did before the disk-wiping incident occurred. But, they would no longer have the convenient place marker (money) through which they previously strained, filtered and accounted for all their productive activities.

The labels rich and poor would no longer have the meaning they previously had. When nobody has any money or debt the definitions attached to the labels rich and poor would shrink in meaning to include physical possessions but, beyond that, would detach from the previous association they had with the label power.

People who have and control money have power. Without money they would, by definition, lose all the power they had, previously attached to the money they had and controlled.

People would still have power but it would quickly shift to other things like physical strength, access to resources, the knowledge and ability to use those resources and the possession of guns and other kinds of weapons.

This thought experiment is interesting because it can unfold in so many different and interesting ways. In a worst-case scenario it's easy to imagine society, at large, devolving into a dystopian nightmare. The strong and well armed would take what they wanted and deal directly and summarily with anyone who got in their way.

At the opposite end of the spectrum of outcomes, everyone would realize that money never was *real* and figure out how to live happily and productively without it.

But a more probable outcome would lie somewhere between the dystopian and utopian extremes. We'd have to reconstruct the financial systems that existed before the great disk-wiping incident and, attempt to, somehow, reallocate the financial assets and liabilities which previously existed.

And even though we are now better and more widely connected than ever before, with observational consciousness exponentially expanded compared to the way it was several decades ago, getting the financial system back up and running—equitably—would be a monumental task.

Individuals and governments would be required to revisit the meaning of the labels rich and poor, fair, just and value.

Of all the labels a situation like the one depicted in this thought experiment would force us to reexamine one of the most important and meaningful would be wealth. And while it is impossible to predict how and to what extent our understanding and appreciation of the label wealth might change, I believe it is reasonable to assume its immediate, automatic and durable association with the label money would change in substantial and meaningful ways; if for no other reason than the truth of Fuller's observation that it is impossible to learn less.

Mind versus brain

What's the difference between mind and brain? This question has been with us since we discovered what the brain is, how it is structured and how it functions. Philosophers of almost every stripe have been wrestling with the mind/brain problem for centuries and, as yet, the issue remains open; no consensus exists which resolves the questions regarding how, if and, to what extent mind and brain are connected, causally or in any other significant way.

Ancient Egyptians didn't believe the brain was very important. We know this because when we examine mummies' tombs, we discover that while great pains were taken to preserve the hearts of mummies, often by placing them in elaborate containers or boxes, mummy's brains were unceremoniously scooped out and discarded. Brains were considered to be insignificant.

Once we discovered what the brain is and how it works, questions about the relationship between brain and mind arose and those questions are as vexing today as they were when we first realized brain, mind and consciousness seem to be inextricably entangled.

Rene Descartes drew a bright and durable line between things physical and things mental. And while he didn't deny the existence of purely mental phenomena or perceptual phenomena like color, taste, and smell he said such phenomena exist in parallel with the physical world but are unimportant and not to be trusted creating the Cartesian dualism most Westerners are aware of and believe in.

For Descartes, the mind or soul is something *extra* that exists in parallel to the physical universe but isn't part of it and, therefore, not to be trusted. Gilbert Ryle, a well-known Oxford philosopher labeled Cartesian dualism "The Ghost in the Machine" deriding Cartesian dualism and accusing those who ascribe to it as being guilty of having committed a category error.

The truth is nobody really knows what, if any, relationship exists between mind and brain. When we focus on the question we become rapidly entangled in issues associated with consciousness, qualia, the existence or non-existence of matter and how material substance, under any circumstances, could possibly give rise to immaterial phenomena like consciousness, love, hate, envy and greed.

Cartesian dualism also created the idea that the physical universe is a closed system meaning that anything outside the system cannot possibly affect anything inside; especially immaterial things like thoughts. But, here is the problem with that. If I want to raise my right arm all I have to do it think about raising it and it goes up. If I see a car speeding toward me, I will get out of the way.

Some believe all mental phenomena are reducible to events occurring in the brain and can be explained by certain identifiable patterns of neuronal firing in areas of the brain that have been studied and mapped like the visual cortex and other areas of the brain associated with things like smell, hearing, taste, and touch. And while other areas of the brain have been identified, associated with phenomena other than perceptions, like music appreciation, mathematical acuity and whether someone is lying or telling the truth, the fundamental question, how can physical processes give rise to non-physical phenomena remains unanswered.

The opposite position which posits mental phenomena as being separate from physical processes going on, in the brain, fails to explain why certain areas of the brain have been proven to be associated with non-physical phenomena like our five senses and our ability to recognize people and events. Individuals who have suffered trauma to areas of their brains associated with non-physical phenomena and sensory input no longer have the capabilities they had before the trauma. Damage to the visual cortex results in blindness and damage to certain other areas of the brain results in the inability to recognize people the person could, before the trauma occurred.

It seems the more we pursue the question the more difficult it is to reach a satisfactory conclusion upon which everyone can agree.

Computers provide rich and sumptuous food for thought regarding the relationship between brain and mind. Brains and computers have striking similarities. They are both physical entities and early computers were often referred to as electronic brains. They both store and retrieve memories. They both process information. They both accept input and produce output. All this raises other questions associated with the mind/brain problem like, if a computer can process information in ways that yield results equivalent to what humans do, is the computer *thinking* in the same sense that a human does? This is the question the discipline of artificial intelligence or AI attempts to answer and it is also an area of philosophical inquiry. And, while computers continue to increase in capacity and speed as their cost decreases, we are still looking for answers regarding whether true or strong artificial intelligence will ever be instantiated in a machine.

When scientists began attempting to create a machine that could "think" they looked for a foil against which to test the capability of the machines they created. The ultimate test of whether a machine can think in the same sense that human beings think is the Touring test developed by Alan Turing. It is relatively simple to understand. Here's how it works.

You set up a machine so it can respond to input from a human being using a terminal and keyboard. The human sits down at the terminal and begins to type. He might type:

"Hello, how are you?" To which the computer might respond. "I'm fine, how are you?"

The point of the Turing test is this:

The human, interacting with the terminal does not know whether he is interacting with a machine or with another human being—also responding using a terminal and keyboard. If he has been communicating with a machine and, if after a sufficient amount of time, cannot determine whether he has been communicating with a human or a machine, the machine has passed the Turing test.

The beauty of the Turing test is the human can ask any kinds of questions he likes and that is what makes it so difficult for any computer. The human can ask questions about things like food, the weather, girlfriends, current events, music, philosophy or religion and the computer must provide responses that seem both appropriate and reasonable. If it cannot, the human will be able to determine whether he is communicating with a machine or with another human being. So far, no machine has passed the Touring test—yet.

In the early years scientists knew the machines available were not robust enough to even attempt the Touring test so they looked for other tasks computers could perform that "required" some semblance of the ability to think. Some examples are the ability to play and win games like tic-tac-toe and checkers and win consistently if the human being they were playing against made an error. They succeeded in doing that but nobody, at the time, believed that ability was a true indication of machine intelligence. As computers became faster, cheaper and more robust, and programming techniques associated with artificial intelligence improved, machine capability also improved. A new foil was identified; the game of chess. Few people would argue that, playing chess successfully requires the ability to think.

Early efforts yielded machines that could play but, not very well so the bar was raised higher; the goal being a machine that could win consistently against a human player. When that goal was reached, the bar was raised again to create a machine that could win consistently against a human capable of playing at the level of master.

On May 11, 1997, Deep Blue, a machine developed by IBM won a six game match by two wins to one with three draws against world chess champion Gary Kasparov.

A machine capable of winning chess against a world renowned chess master is indeed impressive but, the question, of whether Deep Blue is capable of "thinking" in the same sense that Gary Kasparov, or, for that matter, any other human thinks remains open.

The ultimate test remains the Touring test but the bar has been raised again; this time with a much more formidable game which resembles the Touring test but is not quite the real deal—the game of Jeopardy hosted by Alex Trebec on the eponymous television show.

Programmers and scientists from IBM approached the producers of Jeopardy suggesting they pit their skills and machines against the two best players on record who have won the most money playing Jeopardy during the history of the game.

Several key aspects of the way Jeopardy is played present a host of particularly thorny problems for a machine competing against humans. Questions are presented in the form of statements or facts like, "He wrote the famous poem The Raven." To which the answer must be given in the form of a question; "Who is Edgar Allen Poe?" The question categories are broad and include subjects like cinema, geography, authors, puns, plays on words and word combinations, history, travel, current events, food and drink.

Playing successfully, by definition, requires a general knowledge base that is both broad and deep along with the ability to come up with answers which take into account not only the meaning and content of the statements but also the context in which they are presented with reference to the categories, under which, they exist.

During each round, Alex Trebec asks the question in the form of a statement and the three players each have the opportunity to press a button allowing them to "ring in" with their answer in the form of a question. The first player to "ring in" gets to answer the question. If he or she fails the other players get a chance. The game is lively, engaging and difficult because it requires a broad and deep knowledge base along with the ability to understand the questions and how they relate to their assigned categories.

In February of 2011, the machine IBM built and programmed to play Jeopardy, named Watson, played against the two best players in the history of the game. Watson won.

Like the two human players Watson was required to "ring in" after receiving the question by pressing a button just like the ones the human contestants had. But, unlike the human players, Watson received the questions electronically while Alex Trebec read the questions to the human players.

One other important aspect of the game is strategy. Players must choose the category of each question from those displayed on a board they can all see and they must choose the dollar value of the question under the category they chose. If, during the game they randomly land on a question called a "daily double" they're given the opportunity to double the amount of money they can risk on the question. Clearly there is a lot to think about when playing the game of Jeopardy.

Watching a machine win Jeopardy against the best human players who have ever played the game was indeed impressive and it raises the question; does that mean Watson is able to think?

Jeopardy is a different kind of foil compared to chess. Chess is a game of strategy and while it is certainly complicated and difficult, it is also limited to the rules that govern the game in terms of how certain pieces are allowed to move and the layout of the squares on the chessboard. In that sense key aspects of the game are predetermined and cannot be changed.

Not so with Jeopardy; categories and questions appear randomly so contestants have no way of knowing what the questions or categories will be, in advance. Compared to chess that is a very different kind of challenge because, in addition to requiring that players understand the rules of the game they must also understand the content and context associated with categories and questions that appear. From a computer intelligence perspective this is arguably a more difficult challenge than the game of chess.

Watson has access to a huge database and sophisticated programs which allow it not only to retrieve information but also to learn from mistakes it has made while being prepared to play Jeopardy on television in front of millions of viewers. It did not, however, have access to the World Wide Web. So does that mean Watson is able to *think* in the same sense that the human contestants were able to think? And if your answer is yes does that mean Watson is smarter than the humans against whom it played? Or, put differently, is it able to think better and faster than they could?

On the surface, the sheer evidence of Watson's performance appears to prove incontrovertibly that it can not only *think* but that it can indeed think faster and better than its human opponents. But it is important to be careful about jumping to a conclusion so broad and sweeping.

For one thing, Watson has not passed the Touring test. And, while playing Jeopardy *appears* to be very similar to the Touring test it is important to keep in mind that it is *not* the Touring test.

The fundamental question regarding Watson's performance is what was going on in its circuits while it was playing Jeopardy? Was the activity in its circuits similar to the neuronal firing that was going on in the brains of its human opponents? And, if we assume for the sake of argument that it was does that mean Watson possessed the *knowledge* its human opponents possessed? Put differently does Watson truly *know* anything in the same sense that humans *know* things? This is a question with deep philosophical implications.

One way to think about it is to ask questions about what Watson cannot do compared to what its human opponents, or for that matter, any human can.

For one thing Watson is unable to respond appropriately to humor. Watson will not laugh if someone tells it a joke. It also will not know or know how to respond if someone makes fun of one of its answers or derides it for having made an error.

The immediate question this raises is could Watson be programmed to respond appropriately to humor, people making fun of it and derision? If we answer yes then the question becomes are Watson's programmed responses the same as those of a human? Would it be reasonable to assume Watson is experiencing embarrassment or possesses the same kind of *understanding* patterns humans possess in order to be able to see the humor in and be able to laugh at a joke? And if we answer yes, the next obvious question is, does Watson experience the same or similar kinds of *feelings* humans do when they are embarrassed or hear a joke and laugh?

It is here where the issues we are attempting to come to grips with become philosophically murky because they immediately become entangled with deep philosophical questions like; does our improved Watson have a mind? Does it truly *understand* humor? Does it understand, not only what it means to be embarrassed but also, why?

These are questions regarding computer functionalism or, as philosopher John Searle baptized it, strong Artificial Intelligence.[35]

The emergence of computer functionalism was heralded as one of the most important and exciting developments in the entire history of the philosophy of mind in the 20th century. Many believed it was the definitive solution to problems philosophers had been wrestling with for 2,000 years. The idea emerged as a result of the convergence of work that had been ongoing in philosophy, cognitive psychology, linguistics, computer science and artificial intelligence.

"It seemed that we knew the answer to the question that faced us: the way the system works is that the brain is a digital computer and set of programs. We had made the greatest breakthrough in the history of the philosophy of mind: mental states are computational states of the brain. The brain is a computer and the mind is a program or set of programs. A principle that formed the foundation for any number of textbooks was this: the mind is to the brain as the program is to the hardware.

Mind : Brain :: Program : Hardware[36]"

Here, many believed, was the Rosetta stone regarding the relationship between mind and brain; the solution to the problem that so bedeviled Descartes and even vexed the early Greek philosophers from as far back as 2,500 years. In all that time the relationship between mind and body had been a mystery but there was no mystery associated with the relation of a computer program with computer hardware. It is well understood and used daily to program computers.

Does Watson exhibit strong Artificial Intelligence? I don't think anyone really knows but if John Searle's argument against Strong AI is valid, then neither Watson nor any other computer will ever exhibit Strong AI and the questions regarding the mind/body problem and thinking are as difficult, vexing and florid as they've ever been. His argument is a thought experiment called The Chinese Room.

The focus of the attack is a direct appeal to first person experiences when testing any theory that purports to be a theory of mind. Searle asks us to consider the problem backwards by suggesting that if the premise of Strong AI is true, anybody could acquire any cognitive capability by simply implementing the computer program which simulates that capability.

He then suggests we try this with Chinese. First, he makes it clear he does not understand Chinese and cannot even distinguish the difference between Chinese and Japanese writing.

He then asks us to imagine he is locked in a room filled with Chinese symbols. He also has a rulebook which, for the purposes of his argument is effectively a computer program. The rulebook enables him to answer questions put to him in Chinese.

While in the room, he receives symbols which are questions but he doesn't know that they are questions. As far as he can tell they are just symbols. His rulebook tells him what he is supposed to do with the symbols. He picks up other symbols from boxes in the room and manipulates them according to the rules in the rulebook (program) and hands out the required symbols which are interpreted by the person who receives them as answers.

In this scenario we can safely assume that Searle passes the Touring test for understanding Chinese.

What makes this thought experiment so interesting and compelling is we know—as does Searle—that he doesn't understand a word of Chinese. And his conclusion is that if, based on following the rules in the rulebook (program) he doesn't understand Chinese, then neither would any computer understand Chinese simply because it could implement the same program. And the reason he gives for this conclusion is the final nail in the coffin of Strong AI—no computer has anything he doesn't have.

What the Chinese room thought experiment brings into full view is the difference between computation or symbol manipulation (which is what computation is) and true understanding if you imagine what it is like for him to answer questions in English.

You have to imagine he's in the same room but, instead of being handed symbols he knows nothing about, he is given the questions in English which he then answers. From outside the room, Searle's answers to the English and Chinese questions are equally good. He passes the Touring test for both languages. From inside the room however, there is a huge difference.

Here's the difference. In English, Searle understands what the words mean while, in Chinese, he understands nothing. When he's manipulating Chinese symbols, he's just a computer.

Searle's Chinese Room argument went directly for the jugular vein of the Strong AI project. Earlier attempts to attack Strong AI posited that the human mind has certain abilities computers don't have and cannot acquire. The problem with this strategy is it always ends up being dangerous because the moment someone says there is a certain kind of task computers cannot do, programmers, computer scientists and cognitive scientists are sorely tempted to design machines and programs fully capable of performing the task deemed impossible. Early examples are programs and machines that could play tic-tac-toe, checkers, chess at the master's level and, finally Watson which played Jeopardy and won.

Every time those working on Strong AI succeeded in vaulting whatever bar was raised, naysayers said the task wasn't all that important and the computer success simply doesn't count. Those who work on Strong AI justifiably feel that the bar is constantly being raised.

What makes the Chinese Room argument so devastating is it is based on a completely different strategy. Instead of raising the bar, once again, even higher, it assumes those working on Strong AI will completely succeed in simulating human cognition—in designing, building and programming a machine capable of passing the Touring test for understanding Chinese or, for that matter, anything else. But the point is this—regarding human cognition, such achievements are irrelevant; not for a trivial reason but rather for a reason that is *deep* and here it is.

Computers operate by manipulating symbols. The processes they use are purely defined by syntax alone. Put differently what they *do* is what Searle did in his Chinese room thought experiment.

The human mind, however, has much more than just symbols which cannot be interpreted. It attaches meanings to the symbols it has.

Syntax alone is insufficient for providing meaning because syntax is simply the study of the principles and rules for constructing sentences in natural languages. Its focus is on things like grammar and punctuation.

Syntax, in this thought experiment, is the rules in the rulebook or program that Searle used to manipulate the symbols inside the Chinese room. From outside it *looks like* whomever or whatever is handing out answers, understands the *meaning* of the symbols received. But, we know that's not what's happening.

Semantics, unlike syntax, is the study of meaning. The focus of semantics is on the relation between signifiers such as words, phrases, signs and symbols, and what they *mean* or stand for. Semantics is a very different kettle of fish compared to syntax.

Here, again, we run into the *slipperiness* we often encounter when dealing with issues and concepts like consciousness and the speed and composition of light which, on the surface, appear simple and straightforward but, when examined carefully prove to be inherently complex.

If you think about the Chinese room thought experiment carefully, it becomes obvious that Searle assumed there weren't any problems associated with attributing syntax and computation to the system. But, as Searle points out, thinking carefully about the Chinese room reveals something interesting. Computation and syntax are observer relative which means absent an observer who can assign meaning to either or both, by definition, they have no meaning.

As far as we know, there are no intrinsic or original computations in nature except for when someone is performing some sort of computation in his head. If I think to myself 50 + 40 = 90 what's happening in my head is not observer relative because I'm doing the calculation, whether or not anyone knows what I'm doing or thinks anything about what I'm doing.

If I punch in 50 + 40 = on my calculator and it shows me a result of 90, it's a safe bet that my calculator doesn't *know* anything about computation, basic arithmetic, symbols or numbers because, by definition, it knows nothing about anything. A calculator is an electronic device I can *use* to perform calculations. What its circuits do when I punch in numbers are simply electronic state transitions built-in to the device. Any calculation that occurs, as a result of anyone using a calculator is in the eye of the beholder or user. The same applies to anyone using a computer.

The assumption that computation exists in the machine is analogous to the assumption that information exists in books. In both instances the computation and the information are observer relative. Absent the presence of an observer who knows how to use the computer or who knows how to read what's printed on the pages of the book there is no computation or information because neither computation nor information are *intrinsic* to computers or books.

This is the reason it is not possible to discover that the brain is a digital computer. Computation rather than being *discovered* in nature is *assigned* to nature. Asking if the brain is a digital computer is a rather ill defined question.

The answer to the question, is the brain intrinsically a digital computer? Is no because nothing is intrinsically a digital computer other than someone doing arithmetic in her head.

If we ask can we assign computational interpretation to the brain? The answer is yes because we can assign computational interpretation to anything. That's what the term observer relative is all about.

John Searle is a well-known and brilliant philosopher. The ideas and concepts he has put forward regarding artificial intelligence and computation, in general are insightful and thought provoking but they do not, in my opinion, answer the questions about the relationship of mind and brain. The issue is, as yet, unresolved.

On the question of mind versus brain, once again I offer some observations on the subject by R. Buckminster Fuller.

"I am enthusiastic over humanity's extraordinary and sometimes very timely incongruities. If you are in a shipwreck and all the boats are gone, a piano top buoyant enough to keep you afloat that comes along makes a fortuitous life preserver. But this is not to say that the best way to design a life preserver is in the form of a piano top.

I think that we are clinging to a great many piano tops in accepting yesterday's fortuitous contrivings as constituting the only means for solving a given problem. Our brains deal exclusively with special-case experiences. Only our minds are able to discover the generalized principles operating without exception in each and every special-experience case which if detected and mastered will give knowledgeable advantage in all instances."[37]

One of Fuller's favorite examples of a generalized principle was the lever principle. He often pointed out that what makes the lever principle and any other generalized principle so valuable is, that once the mind apprehends it we are able to take advantage of it in a wide variety of circumstances because it always works whether we're applying it outside, in the house, to pull nails for instance, under water or in outer space. A screw, even though it looks nothing like what we normally think of as a lever or inclined plane works on the generalized lever principle. Using a fine enough screw situated between two metallic plates it is possible to exert tons of pressure between the plates using only your thumb and forefinger.

And finally from the transcript of an interview:

> GERSTNER: Should we legitimately fear intelligent machines taking over?
>
> FULLER: Absolutely not. I differentiate between brain and mind. Machines can do what our brains do. The brains of all creatures are always and only recording and integrating the information of our senses—smelling, seeing, touching, hearing. We find then the human mind from time to time discovering a mathematical relationship that is purely intellectual and can only be experienced that way. There's an intellectual integrity operating in the universe. The human mind has some access to the great design of the universe. No other creature has it nor can employ this information objectively. Wow! This means that we humans are here for some very important

> reasons which we have not as yet come to think out publicly. I made the working assumption that we are here as local Universe information-gatherers, and local Universe problem-solvers in support of the integrity of an eternally regenerative Universe.
>
> GERSTNER: So the mind makes the difference?
>
> FULLER: We are completely metaphysical. We're not our body. At my age, I have consumed over 300 tons of food, air and water that became temporarily, my hair, flesh and bones, and I am sure that's not me. We're not physical. What is unique about humans is this mind thing.[38]

For all the work we've done exploring the relationship between mind and brain, at the end of the journey, we find that the *deep* questions remain unanswered. Descartes' duality remains unresolved and, for many, the only plausible answer is the ghost in the machine.

Will machines emerge that can think as well as any human? I believe they will. But, beyond that, I believe machines will eventually emerge capable of thinking orders of magnitude faster and better than all the humans who have ever lived.

The rate of technological change is inherently exponential not linear. When looked at from a narrow perspective, technological change often *appears* linear but what's really happening is the curve is moving toward a *knee* on a graph that charts progress over time. Once the time period reaches beyond the knee, the rate of change becomes exponential and explosive. Computer technology is the best example we have of this mechanism. We are approaching the knee of the curve. When we move beyond it, first strong AI will emerge and then, in short order super intelligent machines will emerge.

The prospect of the emergence of super intelligent machines is both exciting and disconcerting. It is exciting because, with them, we will gain capabilities that could not be possible without them. In a best case scenario they could be relied upon to do all our intellectual *heavy lifting* helping us to deal effectively with issues like war, famine, disease, poverty and distribution of resources.

It is disturbing because the existence of such machines suggests that, once they have emerged, human beings will have found their niche in the grand scheme of the universe. Right now we consider ourselves to be the top dogs. To get an idea of what this means consider this scenario put forward by Edward Friedken from the Massachusetts Institute of Technology where he gives us some idea of how fast super artificial intelligent machines might be able to think.

> Say there are two artificial intelligences, each about the size of a small table. When these machines want to talk to each other, my guess is they'll get right next to each other so they can have a very wide-band communication. You might recognize them as Sam and George, and you'll walk up and knock on Sam and say, 'Hi, Sam. What are you talking about?' What Sam will undoubtedly answer is, 'Things in general,' because there'll be no way for him to tell you. From the first knock until you finish the "t" in about,

> Sam probably will have said to George more utterances than have been uttered by all the people who have ever lived in all their lives. I suspect there will be very little communication between machines and humans, because unless the machines condescend to talk to us about something that interests us, we'll have no communication.
>
> For example, when we train the chimpanzee to use sign language so that he can speak, we discover that he's interested in talking about bananas and food and being tickled and so on. But if you want to talk to him about global disarmament, the chimp isn't interested and there's no way to get him interested. Well, we'll stand in the same relationship to a super artificial intelligence. They won't have much effect on us because we won't be able to talk to each other. If they like the planet and don't want to leave, and they don't want it blown up, they may find it necessary to take our toys away from us, such as our weapons.[39]

On the surface, Friedken's conclusions regarding the implications associated with the emergence of super intelligent machines are somewhat bleak. It is more than, just a bit, disconcerting to be compared to a chimpanzee in terms of intellectual capability when thinking about how we would relate—intellectually—to a super intelligent computer. But, it's important to keep in mind that we have long passed the point where we could design the next generation of computers relying solely on human intellect. The central processing units or CPUs of modern computers are so complex that we *need* computers to continue improving our designs. We need computers to be able to design better computers. And we already have, use and rely on computers and computer programs to help us do things that, without them, we could not do, like making medical diagnoses, solving complex mathematical problems, and searching for oil.

Diagnostic imaging modalities like CT and MRI would be impossible without computers capable of processing the torrents of data these machines produce when used to scan our bodies.

The intellectual niches we once considered to be strictly in the domain of human intelligence have been overtaken by machines we rely on, to the extent that, without them, it would be difficult to survive. Computers have become so ubiquitous that they are imbedded in things we use every day like our automobiles, appliances and entertainment devices.

So, from an appropriately narrow perspective, focusing on niche applications, we are already at the point where comparing our capabilities to what the computers in say a CT scanner can and do already makes us look a bit like a chimpanzee.

But the kind of machines Friedken describes raises a different set of questions with implications that reach far beyond niche applications and extend deep into the domains of ethics, philosophy and myth.

The question most often raised regarding the emergence of a super intelligent machine is would such a machine be conscious? If we interpret that question as being would such a machine *have* consciousness I believe the answer must be no for the same reason that we don't *have* consciousness. We are, instead—in—consciousness.

If we push the question a bit harder by asking, if such a machine could see, hear, smell, taste and touch and then ask would it be—in—consciousness in the same sense humans and other living things are in consciousness the situation becomes much more difficult to deal with.

John Searle would answer no because regardless of what intellectual capabilities a super intelligent computer might possess, at bottom what it is doing is manipulating symbols based on syntax and, even though what it is doing *looks like* bona fide semantic capability, it is not.

When super intelligent machines emerge, we will be forced to deal more deeply with questions regarding whether and to what extent their ability to *think* means the same thing as when we talk about humans being able to think. And, even then, I believe the deep philosophical questions regarding mind versus brain will remain unresolved.

The expansion

Earlier, I posited that observational consciousness is expanding rapidly by noting how Facebook, Twitter, smart phones, the IPad and other web-based technologies make it easier for us to communicate with greater ease and speed than was previously possible. I also said we were witnessing a bona fide spike in the expansion of observational consciousness.

Since I wrote that, we have all witnessed a set of consequential circumstances attached to that spike as the citizens of Egypt, by connecting with each other and the world, forced the ouster of Hosni Mubarak, who had been in power for over thirty years. This is a startling example of how expanded observational consciousness has a direct affect on how we relate to each other, globally, and, of equal importance, how power shifts when observational consciousness expands so rapidly.

What happened in Egypt is affecting the entire Middle East as citizens in neighboring countries like Bahrain, Yemen, Libya and Iran attempt to replicate what the Egyptians did. They have the same access to the technologies the Egyptians had. And, whether or not they succeed, to the extent the Egyptians did, the spike in global observational consciousness is ensuring that the entire world is watching as events unfold, in real time, just as they did during the Egyptian uprising.

The *spike* represents a geometrical expansion in the *flow* of observational consciousness on a level we have never experienced before. Instead of increasing in a linear arithmetic fashion it expanded several orders of magnitude in a very short period. We have approached and are beginning to witness what happens when we begin to pass the *knee* of the curve representing the exponential growth of observational consciousness. It has become explosive and it will continue to not only grow but also accelerate.

Among other things, this spike demonstrates the power that ordinary people now have access to and how effectively they are able to wield that power.

The world has changed irreversibly but the question, "have we become more powerful than we know how to be?" as yet remains unanswered.

What happened in Egypt and is happening across the Middle East is the first concrete example of how the power shift associated with the expansion of observational consciousness is directly affecting the lives of individuals across the globe. Everybody sees what's happening and, whether they want to or not, they will be forced to deal with the repercussions of these events, directly and indirectly.

What makes this situation as worrisome as it is exciting is this is just the beginning of a permanent change in the way we relate to one another. It happened so rapidly that the consequences have taken everyone by surprise. It isn't often that the entire world is surprised so rapidly by events like these. And, part of the problem with what we are witnessing is, few people know what observational consciousness is, that they have access to it, or why it's important.

Language and communication capabilities facilitate observational consciousness and in the past, expansions in our ability to communicate also resulted in expanded levels of observational consciousness. It started with language. Then writing expanded it further. The printing press forced an even greater expansion as did the telegraph, the telephone, radio and television. And, as we adopted and adapted to those technologies, power shifted and our lives were changed in ways we could not possibly have foreseen.

With the exception of television, the impact of previous communications technologies we adopted and adapted to was gentler than what we are now witnessing. They were farther away from the knee of the curve. The changes television wrought on society were faster and more powerful especially in the ways it changed how we gained access to news. But, like the telegraph, telephone and radio, television required a massive corporate infrastructure to make what it had to offer readily available to vast segments of society.

Until recently, the very idea of an individual being able to produce and broadcast a television program lasting even just a few minutes, was too absurd to even consider. It would cost too much and distribution would be, essentially, impossible. That was before You Tube, cell phones that record video, and the internet.

Unlike before, today anyone with a smart phone can record and broadcast events happening where they are, across the globe via the internet making the information available to millions of people instantly. With that capability, even people who cannot understand the audio portion of what was broadcast can see the video and reach conclusions about what it is they are watching. If a picture is worth a thousand words, a streaming video, uploaded from a smart phone, is worth hundreds of thousands.

Wittingly or unwittingly we are all witnessing and participating in the emergence of what Raymond Kurzweil calls the Singularity. The expansion of observational consciousness is the starting point and, like it or not, we are all along for the ride. Computers, artificial intelligence and the internet have played a key role in expanding observational consciousness.

Nanotechnology will have an equally powerful impact on who we are and how we relate to each other and the planet. It will drive an expansion that will have an even more direct impact on not only who we are but what we are as well.

The ultimate goal of nanotechnology is to develop machines the size of molecules able to manipulate individual atoms and computers the size of very large molecules that can direct and control these tiny machines. In the parlance of nanotechnology, these microscopic machines are called assemblers and they will have the ability to make copies of themselves or be self-replicating.

Once nanotechnology advances to the level where we have reliable assemblers and the molecular computers to control them we will have reached the knee of yet another curve associated with Kurzweil's Singularity; a point at which we will see an exponential expansion in our capabilities associated with manufacturing, healthcare and medicine.

We know how to manufacture things and our capabilities in manufacturing have steadily gotten better since the beginning of the industrial revolution. But, we haven't yet seen the exponential expansion in manufacturing capability that we have with computer technology and electronics in general. And even though we now have the capability to build chips with as many as a billion components compared to nanotechnology using assemblers that capability is primitive.

We know how to manufacture automobiles. The manufacturing process is layered and complex requiring resources like steel, aluminum, glass, rubber and plastic. Individual components are processed and manufactured in separate facilities and ultimately brought together in a plant designed to facilitate assembly of the individual components into the finished product. It is a labor intensive and expensive process but has, nonetheless, resulted in a wide variety of automobiles that have consistently improved over time. Most people are familiar with the processes associated with automobile manufacturing.

With nanotechnology the process would be very different, so different in fact that it will bear no resemblance to how we do it today. Here is how it might work.

Imagine a vat large enough to contain a finished automobile. Inside that vat there is a large plate on which the finished automobile will sit. In the center of that plate is a large molecule sized computer, too small to see with the naked eye, which contains all the instructions required to build an automobile atom by atom. Now imagine that the vat has a large viewing window allowing you to see what happens inside.

In the first step of the process the vat fills with a milky fluid. This fluid contains the assemblers. Imagine that the computer is shaped like a dodecahedron. As the assemblers come in contact with it they attach to one of the facets and extend their tiny arms outward allowing other assemblers to grab their extended arms. The number of assemblers required is programmed into the computer sitting in the center of the plate. When the required number of assemblers has been extracted the milky fluid is evacuated from the vat. What's left behind looks like a gossamer bubble shaped something like a car.

In the next step of the process the vat fills again, this time with a fluid which is an element slurry and also contains the fuel the assemblers need to do their work. This slurry contains the atoms needed to make tires and wheels. As the required atoms float past the assemblers they grab the ones they need and assemble them, directed by the computer, into the tires and wheels of the car. When that task is complete, the vat empties and refills with the elements required to assemble the frame chassis and suspension components of the car, once again, atom by atom. This process continues until every component of the car has been assembled from its wheels, tires, frame, chassis, suspension, engine, body, glass and paint to all the electronic components required for the car to run and function properly. At the end of the process the assemblers and element slurries are evacuated and we're left with a completed automobile which is extracted from the vat.

This process is very different from how cars are assembled now. With nanotechnology, manufacturing a car is more like growing a tree than manufacturing and assembling individual components. In areas where the car must be soft, like the seats and the rubber in the tires it is soft. In areas where it must be hard, like the suspension, chassis, wheels and engine it is hard. There are no seams or rivets or welds because it was assembled atom by atom and, in that sense, is more of an organic whole rather than an assembly of disparate parts.

No people were directly involved in the assembly of the car. No steel had to be smelted, no rubber processed, no glass shaped and molded, no body panels hammered and bent into shape and no paint booths to spray on the color. Every component of the car from the fabrics for the upholstery to fuel injectors was assembled atom by atom.

It may be difficult to imagine that we could develop the technology required to manufacture cars as I've just described but it's important to remember that this kind of *assembly* goes on in every cell in our bodies and in the bodies of all living things. Our cells assemble the proteins they need, when they need them directed by the DNA in their nuclei.

The medical applications of nanotechnology are even more mind boggling than manufacturing. Assemblers could be designed to be injected into our bloodstream to seek out tumors and disassemble them, atom by atom, and replace the disassembled structures with healthy tissue assembled atom by atom. They could also be designed to circulate continuously in our bloodstream to monitor and repair the effects of aging allowing us to choose what age we would like to be and stay at that age. And for older individuals, the same technology could be used to reverse the effects of aging bringing them back to the age they would like to be at and maintain.

With this technology, people would be able to live as long as they wished without having to worry about the ravages of age and disease.

These capabilities will be part of the expansion or what Ray Kurzweil calls the Singularity.

Nanotechnology, when it emerges as I have described it, will forever change the ways in which we relate to each other and to the planet. And, while at first blush, it seems so fantastic as to be out of the question, it is important to keep in mind that Mother Nature has been using it since life emerged to create the overwhelming diversity we see in all living things from the lowly amoeba to geniuses like Albert Einstein. Every cell in our bodies is replaced approximately every seven years and the processes required in replacing tissues such as bone, muscle, organs, and tendons are very much like the ones I described in the atom-by-atom assembly of a car in a vat using assemblers.

The *replicators* depicted in Star Trek gave us our first glimpse of what we can expect when nanotechnology emerges.

Beyond being able to manufacture things we are already familiar with it will allow us to create materials with properties that cannot exist using existing manufacturing techniques. Once we gain the ability to assemble things atom-by-atom, we will be able to control the intrinsic properties of the things we make yielding, for instance, plastics that are lighter and stronger than steel.

Along with holding out the prospect of being able to live as long as we'd like to live, nanotechnology will also make it possible, for those who might be interested in such things, to enhance our ability to think. As impressive as our brains are, in terms of form, function and numbers of neuronal connections, that same form and function limits their capability when compared to what silicon-based computers can do. Computers can store, index and retrieve tremendous amounts of information; much more than any human can. They can also perform calculations at blistering rates approaching a trillion floating-point operations or FLOPS per second. No human brain could ever compete—unless it was enhanced in some way. Nanotechnology will allow enhancement of our brains ultimately making it possible for humans to think as fast as and access as much information as computers can.

All this raises a host of deep and difficult philosophical issues that we will need to confront and wrestle with as we approach the singularity. As I think about enhanced human brains capable of thinking as fast as computers, I'm reminded of a chilling line from a Pink Floyd song—"There's someone in my head but it's not me."

I have touched only lightly on some of the ideas Ray Kurzweil presents in his book *The Singularity is Near.* If you would like to explore these ideas in greater depth and detail read the book. I found it startling, fascinating and more than just a bit frightening; not because I don't believe any of it, but because I do.

MODELING THE DIMENSIONAL STRUCTURE OF CONSCIOUSNESS

There is a deep and meaningful relationship between structure and function. The dimensional structure of consciousness provides for and enables us and other observers to have the functional experiences we do in the middle level of reality or what we consider everyday existence; the world of tables and chairs, hot dogs, baseball, apple pie and Chevrolets.

We have carefully examined and considered how the pieces fit together and work but the tools we have to work with in attempting to understand what is happening, on an intellectual level, require us to *add in* experiences we have become accustomed to like the apparent existence of space and time which, on the most fundamental level of existence—the quantum level—do not exist.

Let us look first at the functional aspects of consciousness as we have come to understand them within the framework of the dimensional structure of consciousness and, when finished with that exercise move on to a model of the structure we can wrap our minds around using concepts from the mid level of reality where we spend most of our time.

The functions are attached to the realms or domains of consciousness and each realm *is* the function we are able to experience as a result of being embedded *in* consciousness.

Light is visual consciousness. It allows us to see. When we see something we do that as multicellular organisms. Individual cells, including retinal cells do not see anything. For them space does not exist. If we strip away the self or "I" from the function vision or seeing what we're left with is no *seer* and nothing *seen*; all that remains is the perceptual realm of *seeing*.

Sound triggers aural consciousness. *Hearing* is aural consciousness. When moving air molecules make an impact on our eardrums we hear. We experience that as multicellular organisms. Individual cells, including the hair cells in the cochleae of our inner ears do not *hear* anything. For them sound does not exist. If we strip away the self or "I" from the function of hearing what we're left with is no *hearer* and nothing *heard*; all that remains is the perceptual realm of *Hearing*.

Smell is olfactory consciousness. When appropriately shaped molecules come in contact with olfactory receptors we experience smell. We do that as multicellular organisms. Individual olfactory receptor cells do not smell anything. For them, smell does not exist. If we strip away the self or "I" from the function of smelling what we're left with is no *smeller* and nothing *smelled*; all that remains is the olfactory perceptual realm we recognize as *Smell*.

For multicellular animals and humans, seeing, hearing and smelling are their spatial world structured on the visual, aural and olfactory perceptual realms. It is these perceptual realms that open up or create the potential for multicellular organisms to experience seeing, hearing and smelling.

Taste is what multicellular beings experience as the chemical realm of perceptual consciousness. When appropriately shaped molecules come in contact with taste receptor cells on our tongues and in our throats and gut, we experience taste. Individual cells, including taste receptor cells can discriminate between the different kinds of molecules that trigger taste in multicellular organisms but they do not taste anything in the same sense that multicellular beings experience taste. We experience taste because those taste receptor cells send signals to our brains. Taste is a multicellular *chemical* experience. If we strip away the self or "I" from the function of tasting what we're left with is no *taster* and nothing *tasted*; all that remains is the chemical realm of perception we recognize as *Taste*.

Touch is what multicellular beings experience as the tactile realm of perceptual consciousness. When something or someone comes in contact with tactile receptor cells in our bodies we feel a touch or a pain because those cells send signals to our brains. When we experience acceleration or find ourselves in a gravitational field we experience the tactile realm of consciousness. Individual tactile receptor cells can *react* to touch but they do not *feel* in the same sense that we, as multicellular beings, feel touch, pain or acceleration. If we strip away the self or "I" from the function of feeling what we're left with is no *feeler* and nothing *felt*; all that remains is The tactile realm of perception we recognize as *Touch*.

For individual cells there is no space associated with the chemical and tactile or *chemo-tactile* realms of consciousness. Chemo-tactile consciousness is temporal. Only multicellular organisms with access to the visual, aural and olfactory realms can experience taste and touch in a spatial context because for them, space unfolds in the touch of light. Absent vision a chemo-tactile organism that could smell would experience a *kind of* space much less well defined than possible when coordinated with vision. The same would be true for a chemo-tactile organism that could hear but could not smell or see. A *kind of* space would arise but, uncoordinated with vision, it would be ill defined.

The world of individual cells and organisms like plants made up of cells that are purely chemo-tactile is only temporal. For them, space does not exist; not even a *kind of* space.

For individual cells and multicellular organisms like plants, taste and touch are their temporal world structured on the chemical and tactile realms of consciousness. It is these purely temporal perceptual realms that open up or create the potential for individual cells and multicellular organisms like plants to experience taste and touch but their experience of taste and touch is very different from what multicellular organisms who can also see, smell and hear experience as taste and touch.

Because we have access to the visual realm of consciousness, we are able to envision things that do not exist or that we cannot see. When I hear the squirrel chittering in my back yard I envision the squirrel. When physicists try to come to grips with and understand sub atomic particles they envision them as very tiny billiard balls or spheres.

Let us do the same with the dimensional structure of consciousness.

I envision my pattern integrity enshrouded by five concentric spheres with my pattern integrity (not my *self* or "I") at the center. Each sphere is a realm of perceptual consciousness. The central sphere is taste which is enshrouded by the next sphere which is touch. I chose this order because, like all multicellular organisms I started out as a single cell searching for another single cell. After they merged the new cell was still purely chemo-tactile. The other three spheres representing the visual, aural and olfactory realms did not exist for me until I became fully developed. Once born, I could experience all the perceptual realms of consciousness but I could not, yet, experience the observational realm. For that I had to wait until I acquired language.

As I began to understand that others could speak and eventually began to acquire my own ability to speak, the observational realm of consciousness began to emerge.

For the model I envision the observational realm as a sort of cloud which enshrouds the five concentric spheres of perceptual consciousness because the observational realm is made up of perceptual parts and, given the right circumstances, grows over time driven by language, communication, writing, and other communication and information based events and technologies. Its dimensional correlate is order.

My *self* or "I" arises in order to do. Being does not require the *self* or "I". It is my pattern integrity that, during meditation, *watches* my *self* and becomes aware of when it lets go or detaches. It is my pattern integrity that experiences being without doing.

The *self* is very strong. The *self* gathers and maintains labels and attaches labels to perceptual experiences. The *self* envisions the squirrel when it hears the chittering sound by attaching the label squirrel to the sound. The *self* wants to *do*. It wants to go to the window so I can see the squirrel. Without my *self* the chittering sound is not a squirrel. There is no label and the chittering is just a sound in the aural realm of consciousness. Without the self there is no one to hear and nothing heard; all that exists is consciousness.

This model consisting of concentric spheres with your pattern integrity embedded in the center of the centermost sphere is deceptive because I did what I said I would. I created it using dimensional concepts from the middle level of reality where we experience space–time and the physical dimensions associated with space–time; I added them in. It *presupposes* the existence of space–time and, one of the first questions that arise regarding the model is how big the spheres are?

One way to think about that is to ask how big a sphere is. If we focus on that question we can pull ourselves out of the loop that keeps sending us back to the middle level of reality because the answer is it does not matter. The word sphere describes a concept that our mind *sees* as a fundamental geometric solid. Size, color, consistency, mass or what it is made of do not matter. A sphere just *is* regardless of how large or small it might be. Close your eyes and envision a sphere. How big was the sphere you envisioned?

We run into similar problems with light. We talk about light moving through space. We talk about its velocity and momentum. We envision it streaking across the depths of space or across the room and touching our retinal cells and we find ourselves, once again, trapped in the same loop. Where were the photons before they streaked across interstellar space or the space in the room? Where were our bodies and brains and retinal cells, before the photons touched our retinal cells? And it is here where we can begin to *see* the loop that space–time creates when we attempt to envision things that, by definition, *require* a different kind of *seeing*; the kind of *seeing* that can only be done with our minds.

We cannot touch the sphere we envision with our eyes closed. When we think about these kinds of concepts it is the circularity of our reasoning that creates the loop. If we *look* at the loop we can *see* the circularity by reminding ourselves that in our attempt to explain where the photons were and where we were before they touched our retinal cells we *assumed* the existence of space. That assumption is the pivot point, or nexus, of the loop. It leads us to a mental picture that cannot be used to understand what is happening because what is going on is happening on the quantum level of reality and, on the quantum level there is no space. There is no *out there*.

We experience the loop because all we have are our dimensional experiences which exist one click above the quantum level of reality; the middle level of reality. It is here where photons *appear* to move through a space that *appears* to exist and that we apprehend as *out there*.

We are caught in the loop because, we are trying to do something that cannot be done; envision the space through which the photons travel on the quantum level where they actually exist. Our dimensional concepts get in the way and create the loop because we try to add something that does not exist, on the quantum level—space—to the model we are trying to use to understand what is happening.

Space dimensions exist *only* above the quantum level of reality. The loop is revealed.

Space arises when—and only when—we experience photons. It is our *experience* that allows space–time to come into being or arise. The universe unfolds in the touch of light. But a naked photon is nothing more than a point of light that has a relationship to other points of light. It is never really *in* space. It is instead a point of light with a specific frequency and wavelength which we experience as color that *creates* space in relation to other photons we can see. Space–time is in light. The loop arises when we assume, instead, that light is in space–time. And, once again, the loop is revealed.

MIND TRAPS

Our minds are tenacious. They tend to hold on to ideas and concepts which were formed with information received via the perceptual and observational realms of consciousness. It took centuries for us to understand and eventually come to believe that the earth is not flat. Myth is what invites and, often, forces us to hold on to concepts which eventually get proven to be wrong.

Before we invented writing and became literate, information was impossibly evanescent. And before we invented language the only methods available for exchanging information were grunts and gesticulation. We interacted with each other the same way animals do. You can gain a deep understanding of what this means by watching the movie *Quest for Fire* which so accurately depicts what life was like 80,000 years ago it is used in college and university anthropology classes across the globe.

Before we had language, if someone made a discovery like a source of food or water he or she had no way to tell anyone else other than to show him or her physically what was discovered.

Today, in the twenty-first century, most people take language and writing for granted—a given. Children growing up today think nothing of being able to Google something, accessing a database which is huge, far-flung and constantly growing. And Google is just one such database. But, it wasn't always so.

> "Try To Imagine," proposed Walter J. Ong, Jesuit priest, philosopher, and cultural historian, "a culture where no one has ever 'looked up' anything." To subtract the technologies of information internalized over two millennia requires a leap of imagination backward into a forgotten past. The technology hardest to erase from our minds is the first of all: writing. This arises at the very dawn of history, as it must, because history begins with writing. The pastness of the past depends on it.

> It takes a few thousand years for this mapping of language onto a system of signs [writing] to become second nature, and then there is no return to naiveté. Forgotten is the time when our very awareness of words came from *seeing* them. "In a primary oral culture," as Ong noted, the expression "to look up something" is an empty phrase: it would have no conceivable meaning.
>
> Without writing, words as such have no visual presence, even when the objects they represent are visual. They are sounds. You might "call" them back—"recall" them. But there is nowhere to "look" for them. They have no focus and no trace.[40]

That is a startling quote. Preliterate people could not *look up* anything, because they could not read, and there was no place where anything could be looked up. The moment anyone said anything, what he or she said was gone. The air molecules that vibrated when they spoke ceased vibrating and the information contained in what they said, while it could be recalled by those who could remember what they heard, could not be *looked up* by anyone anywhere.

The idea of history and the pastness of the past being dependent on writing creates, for me, another loop suggesting that time, and all the concepts that orbit and surround it, too, depend on writing. Absent literacy, even with language, existence must have seemed to be nothing but a series of constantly unfolding events. Day and night would come and go. Seasons would come and go. Weather would change but the very concepts of both past and future did not exist in the sense we understand them.

Our minds grab on to thoughts ideas and concepts to help us understand ourselves, the world, and how we relate to it. Today we take the idea of history as something not only meaningful but something that has always existed. One reason we do this is the written word exists for us. We use it not only to gain knowledge but, also, to gain and store knowledge about what we know. We use the written word to organize thought.

"We may wish to understand the rise of literacy both historically and logically, but history and logic are themselves the products of literate thought."[41]

This is a mind trap created by our inability to recall a preliterate past. Before we developed writing, access to the image screen and observational consciousness was limited *only* to the spoken word. History, as we understand it now, did not exist.

It is as difficult to imagine what it was like when we were preliterate as it is to imagine what it would be like to be a single cell or to be alive without the ability to see, hear, smell, feel or taste anything.

If you are reading this you are among the many who take literacy for granted. But writing and literacy are much more than simply the ability to communicate across space, time and generations using various types of media. With writing and literacy, in modern society, the medium becomes the message and the message helps us create and shape and give to reality a temporal direction; from past to future.

"Writing, as a technology, requires premeditation and special art. Language is not a technology no matter how well developed and efficacious. It is not best seen as something separate from mind; *it is what the mind does*"[42] (emphasis mine).

That is an extremely important and perspicuous observation: language is what the mind does. Language is the mechanism the mind uses to map words onto objects and, with literacy to map letters onto sounds and meaning onto written words. With literacy the symbols and words become detached from the objects and concepts they represent. They begin to occupy the dimension associated with observational consciousness—order. But, more than just that, the processes the mind uses to enable writing creates order. Order arises as a construct of the mind facilitated by language and writing.

And now, we have become more deeply embedded in the written exchange of words and messages than ever before. We are shaped by the technologies we adopt and adapt to because, when we add them to our lives and lifestyles, they not only reshape reality they create new realities.

Sixty years ago, television shows like "Father Knows Best" and 'Leave it to Beaver" created a powerful mind trap for American society. Those shows depicted an idyllic reality which flowed effortlessly onto our collective image screens through the medium of television which, at the time, was as ubiquitous as the web and web-enabled smart phones are today. They were entertaining and amassed huge audiences. The mind trap was that the reality they depicted didn't match the reality of most of the people who watched the shows; especially those who lived in poverty or the ghetto. Viewers' expectations of what their lives *should* be like rarely matched what they saw on "Father Knows Best" and "Leave it to Beaver," creating a pivot point for disappointment and angst which, while not obvious at the time, defined the shape of the mind trap.

Twenty years ago, if not communicating face-to-face, or by letter or email, we relied primarily on the telephone and voice contact. We used the telephone to reach out and touch someone and often took comfort and pleasure in being able to hear what they had to say and appreciate the nuance only the voice can overlay on conversation. We were able to *hear* when someone was upset, angry, sad or depressed by the tone of his or her voice.

As cell phones began to proliferate, voice contact expanded exponentially. We didn't have to be tethered to a landline anymore to be able to call someone and have a conversation, even if who we were calling was using a landline. Observational consciousness began to expand and our collective reality was reshaped by our expanded ability to have phone conversations with greater speed and convenience.

The emergence of the web-enabled smart phone has had a much more powerful impact on how we relate to one another than the cell phone which only enabled voice contact. The ability to send and receive text messages has reshaped the realities of everyday existence across every spectrum of society and has changed relationships between siblings and parents, husbands and wives, business partners, criminals and the law.

We take this ability to quickly and easily send and receive text messages as part of the new reality created by the availability and ubiquity of web enabled smart phones. But we rarely notice or think about the extent to which this new capability has reshaped our reality and the ways we now relate to each other. Here's an example from *Alone Together* by Sherry Turkle.

> Dan, a law professor in his mid-fifties, explains that he never "interrupts" his colleagues at work. He does not call; he does not ask to see them, he says, "They might be working, doing something. It might be a bad time." I ask him if this behavior is new. He says, "Oh yes, we used to hang out. It was nice." He reconciles his view that once collegial behavior now constitutes interruption by saying, "People are busier now." But then he pauses and corrects himself. "I'm not being completely honest here: it's also that *I* don't want to talk to people now. *I* don't want to be interrupted. I think I should want to, it would be nice, but it is easier to deal with people on my BlackBerry."[43]

And it is here, in this exchange between Dan the law professor and Ms. Turkle we begin to see how powerfully web-enabled smart phones have reshaped who we are, how we relate to each other and our environments. Even ten years ago, we would rarely hesitate to call or drop in on a colleague at work and often, when we dropped in or they dropped in on us we would *hang out* for a while.

We liked the voice time and face time; even reveled in it. But the smart phone has changed all that. It's easier and more convenient to send a text and what was once viewed as normal and pleasant—"it was nice"—has suddenly become politically incorrect and taboo. The shape of observational consciousness has changed becoming both richer and more expansive while simultaneously creating distance and separation and thinness in relationships and the way(s) we interact.

Texting has become a mind trap which, on the one hand, expands and improves our ability to communicate and, on the other, introduces separation and taboos which previously were seen as being *nice*. Our expectations regarding what's appropriate when making contact with each other have been turned inside out. Dan is, after all, in his mid fifties.

Certainly it is the availability of the smart phone and the ability to send and receive texts which has helped create the space in which we see the changes in how we relate. But I believe the changes themselves were wrought by the expansion and change in shape of observational consciousness the smart phone made possible.

We become what we behold and find ourselves changed. Our ability to communicate with each other has expanded, by becoming broader and faster than it has ever been but it has also become flatter because texting strips away the nuance associated with voice contact using the telephone. Compared to face-to-face conversation filled with not only voice cues but also facial expression, body language and gesticulation, texting and email are virtually bereft of nuance.

Texting allows us to establish and maintain constant contact with one another, but it also increases the psychic and emotional distance between those who are texting by flattening the communication space, making it thin—almost gossamer.

This is another mind trap associated not only with the expansion of observational consciousness but also by the speed of that expansion, and here is the reason:

> An e-mail or text seems to have been always on its way to the trash. These days, a continuous stream of text becomes a way of life, we may say less to each other because we imagine that what we say is almost already a throwaway. Texts, by nature telegraphic, can certainly be emotional, insightful, and sexy. They can lift us up. They can make us feel understood, desired, and supported. But they are not a place to deeply understand a problem or explain a complicated situation. They are momentum. They fill a moment.[44]

It is the *momentum* and *throwaway* nature of texts and e-mails that flattens the communication space making it gossamer and insubstantial by stripping value from the communication and creating a space for a new kind of mind trap.

What we gain from this exponential expansion of observational consciousness, speed, convenience and momentum is counterpoised by what we lose in time available for deliberation and thought and the nuance embedded in voice and face-to-face communication.

Fifty years ago, when adolescents began to separate from their parents, the process was substantially different than it is now. Moving from adolescence to young adulthood was a transformative rite of passage leading, ultimately, to true independence.

Today, children as young as nine years of age get their first cell phones. But, when their parents give it to them, there is an implied contractual obligation to answer their parents' calls. This opens a space for the child to do things like going to the mall or the beach or going shopping with friends that they wouldn't be allowed to do without the phone. This electronic tether, while creating a space for freedom simultaneously deprives the child of the experience of being *alone* with nobody to count on but himself or herself. It substantially changes the nature and shape of the passage from adolescence to young adulthood and independence.

With a cell phone and your parents on speed dial, rites of passage that were once exciting and fraught with potential danger, like navigating the city on your own for the first time, become insulated and buffered. Before cell phones became ubiquitous, if an adolescent during a first time city navigating passage became frightened, he or she would have to experience those feelings alone—without an electronic tether.

And, here again, we see how expanded observational consciousness changes expectations and relationships, this time between children and parents, by reshaping the space in which they exist and operate making it wider and more expansive while, simultaneously, making it flatter, safer and less threatening.

We have collectively run down the rabbit hole and embraced another mind trap that both giveth and taketh away.

This rabbit hole is so deep and its impact on us so insidious that people feel compelled to send and receive text messages while driving. Few things, other than driving while drunk, pose more danger to society at large than sending or receiving text messages on a smart phone while driving. People know—instinctively—that it is both wrong and dangerous to text while driving yet this particular mind trap is so strong that cities, towns and villages are compelled to set up police patrols focused on catching individuals engaged in texting while driving to issue citations with substantial fines attached in an effort to subdue the practice.

The smart phone by providing the ability to send and receive text messages and video is reshaping observational consciousness, on a global scale, forcing exponential expansion and rapid maturation. And, through the process, as was the case when the dimensional realms of vision, hearing, smell, taste and touch associated with space and time and mass, expanded, matured and acquired the dimensional structure associated with the niches they occupy in perceptual consciousness, we are now witnessing and participating in the same mechanism, as it applies to the observational realm of consciousness. The realm of observational consciousness is not only expanding; it is also changing its shape and becoming more firmly established as a bona fide niche in the overall dimensional structure of consciousness. And, while it is and always will be made up of *perceptual* parts, it is separating from the perceptual realms of consciousness in the same way the perceptual realms separated from each other. It is becoming better defined and more deeply embedded in who we are, what we are and how we live our lives. The image screen is becoming less evanescent and gaining substance which, eventually, will rival the *substance* we now associate with the quantum and photon screens. We don't see it happening because we are actively engaged in *making* it happen.

This rapid and inexorable expansion of observational consciousness is binding us together, on a global scale, more tightly than ever before. Whether we like it or not; whether we want to or not we will have to find ways to deal with what is happening that preserve rather than threaten our safety—like texting while driving does. We have let the genie out of the bottle and there is no way to get it back in.

The Persistent Illusion of Smoothness

Like stillness, smoothness is an illusion. Until the very beginning of the twentieth century, scientists believed that energy flowed smoothly. When something got hot, like a piece of iron, the flow of energy from the heat source was assumed to be smooth. As the iron got hotter it began to glow, first red, then orange and, if heated long and hot enough it began to glow white. The transition *looked like* it was smooth leading physicists to believe that it was. But, when Max Planck began to investigate black body radiation in his search for materials suitable for incandescent light bulb filaments, he discovered that energy was anything but smooth. Careful examination of the dynamics associated with black body radiation revealed that energy exists in discrete packets which he identified as quanta. His discovery opened the door to the enigmatic realm of quantum mechanics.

Objects that appear to be smooth, like those made of glass or highly polished metal are, like all other material objects, quantized. At the most fundamental level they are composed of atoms which are composed protons, neutrons and electrons entangled in an incessant dance of furious motion.

When you look around, space appears smooth because light appears smooth. But it is not smooth. It consists of discrete bundles or quanta or bits of energy called photons. But, because they are so small and *appear to* move so fast light appears to be smooth. The space they are *in* is the grid they occupy—that grid *is* the photon screen.

Objects we perceive appear to be smooth because, as Samuel Avery points out, our experience is much like looking at a newspaper photograph. If we examine the photograph closely we see, immediately, that it is made up of a series of closely spaced dots of varying density. Viewed from an appropriate distance the photograph looks smooth but it is not smooth.

Photographs taken with film appear smoother than newspaper photographs but that's because they are composed of silver halide crystals which are much smaller than the dots in a newspaper photograph. If you examine a film image with the appropriate amount of magnification you will see very tiny dots of varying density creating an image in the same way the larger dots do in the newspaper photograph. They are both quantized. The difference is in the size of the pixels or bits which make up the image. Film images look smoother than newspaper images because the bits they are made up of are smaller.

Photons of the appropriate wavelength are the smallest bits or quanta we can see. Space–time is in light which is the reason space–time arises when photons touch our retinal cells. The universe unfolds in the touch of light which leads, inexorably to the conclusion that, instead of being smooth, the universe is quantized. This conclusion is based on the physics as we have come to understand it.

What we see happening around us—everything—is painted and sustained by visible light quanta—individual packets or bits of energy that appear to be smooth but are, in reality, discrete quantum level entities on a three dimensional grid we recognize as space. This *grid* is the photon screen; not the photons. When you look around, the space that makes up the grid is mostly empty. Empty space is potential perception which can be filled with definite perception.

Ever since Albert Einstein did his Nobel Prize winning experiment which proved photons exist physicists have been doing experiments with photons, some of which we have already explored in the form of double slit experiments to determine whether the photons will behave as discrete packets or bundles or bits of energy or as continuous waves which can interfere with each other. And, as we have also seen, the results of those experiments defy common sense.

If the photons are released, one at a time, toward the double slits, common sense tells us each photon must choose the slit either on its right or its left before hitting the detector on the other side of the double slits. So what we should see, after enough photons have passed through the slits, is an image of two separate slits on the detector. But, since in doing this kind of experiment, we are interacting with quantum level entities—photons—common sense fails us. We expect the photons to behave the same way tiny billiard balls would behave if we fired them at the slits. And even though we know that each photon we fire at the slits is a discreet albeit quantum level entity, how the photons behave after going through the slits depends entirely on how much information we have or can potentially have after they have passed through the slits.

It we set up the experiment so that we cannot possibly know which slit the photons chose then, instead of an image of the slits or two separate splotches of light we see an interference pattern even though we released the photons towards the slits, one at a time.

If we put a detector in the path the photons take after reaching the slits which allows us to know which slit the photons choose then the interference pattern disappears and we see the two splotches we expected when we began the experiment.

The conclusion about all this is if we set up the experiment in such a way that it is impossible for us to know which slit the individual photons choose then even though they are being released individually, each photon is going through the slit on the right, and the slit on the left and interfering with itself before hitting the detector on the opposite side of the double slits. An even more bizarre conclusion is each photon conforms to what Richard Feynman identified as the sum of all paths. The photon goes through the right slit and the left slit, it goes through only the right slit and through only the left slit and it goes through neither slit yet ends up on the other side of the plate with the slits and interferes with itself before hitting the detector on the other side of the slits. Put differently each single photon takes every conceivable path to and through the slits interfering with itself and painting an interference pattern on the detector on the other side of the slits, hence the term *sum of all paths*.

An interesting variation of this experiment called the delayed choice or quantum eraser variation involves using polarizing film in front of the slits. Polarizing film will cause the photons that pass through it to assume a directional orientation, for instance each photon can be polarized vertically or horizontally or in some direction between horizontal and vertical.

If the polarizing film on the left slit gives the photons that pass through it a vertical orientation and the polarizing film on the right slit polarizes the photons that pass through it horizontally then, after passing through the slits, the interference pattern disappears and we see, instead, two splotches. This isn't surprising because, by polarizing the photons, we gain information about their orientation allowing us to *know* which slit the photons went through.

But, here's where things get interesting and weird. If we leave the experimental set up unchanged but insert another polarizing film in front of the slits, which polarizes the photons from both slits at a 45-degree angle (half way between horizontal and vertical) then the two splotches, disappear and, once again, we see an interference pattern!

If you think about this carefully you will find yourself even deeper down the quantum weirdness rabbit hole than when thinking about the double slit experiment without the polarizing filters.

When we put the vertical/horizontal polarizing filters over the slits the photons *know* that we now *know* which slit they went through and the interference pattern disappears. But, when we insert the 45-degree polarizer into the apparatus after the photons have passed through the slits polarized vertically and horizontally the interference pattern reappears because the photons *know* that we can no longer *know* which slit they went through—after they have passed through the slits!

Yet another version of this experiment has been done using an elaborate setup that splits the photons passing through the slits into entangled pairs which can be made to follow certain paths in the experimental setup. Since we know that if we do something to one photon of an entangled pair we also know what happened to the other photon immediately; it is possible to manipulate the paths of the entangled photons introducing a measurable delay in their arrival at the detectors used to determine what they have done. In the final analysis, even though everything that happens to the photons as they move through the apparatus is determined by the laws of quantum mechanics, what we discover is entirely dependent on what information we choose to *know* about even after the photons passed through the slits.

This delayed choice/quantum eraser version of the double slit experiment was proposed by one of the giants in physics, John Archibald Wheeler in 1978. At the time he proposed it we didn't have the apparatus required to actually carry out the experiment. It was just a thought experiment. In the early 1990s the experiment was actually carried out using electrons and proved that as unintuitive as it may seem, the electrons seem to *know* what choice the experimenter will make about what paths they took after passing through the slits even when time delays are purposely introduced into the experimental setup after detection has occurred earlier in the setup.

The bottom line in all this is: what the experimenter *sees* at the end of the experiment is entirely dependent on what he chooses to *know* about what the photons or electrons did, in the apparatus, even after they have already been detected at an earlier time before the experimenter decided what question he was going to ask. WTF indeed!

John Wheeler's ideas and thoughts about the universe and how it works were big, bold and all encompassing. He coined the cosmological term black hole which describes an object; in the cosmos with gravity so strong that even light cannot escape. We cannot *see* a black hole because light cannot escape from a black hole but, because its gravity is so great we can determine and infer its presence through astronomical observation. We know that black holes exist and, most astronomers believe gigantic black holes exist at the center of every galaxy including our own, the Milky Way galaxy.

In discussing the delayed choice or quantum eraser version of the double slit experiment and the seemingly bizarre and unintuitive results physicists get when they conduct the experiment, Wheeler compared what they were doing to playing the surprise variation of the game 20 questions.

In the surprise variation of the game, one player leaves the room while the rest of the players—or so the person who left the room believes—selects some person place or thing. The player who left then comes back into the room and attempts to guess what person place or thing the others have chosen by asking a series of questions that can only be answered yes or no.

What this player does not know is the group has agreed to play a trick. The first player to be questioned by the player who left the room will think of an object only—after—the question has been asked. Each player, in turn, does the same thing, ensuring a response that is consistent not only with the immediate question but also with all the pervious questions.

As Wheeler points out, what's important about all this is the word *wasn't* in the room when the player who left, came back into the room, even though he thought it was.

In relating this game to what happens in the delayed choice version of the double slit experiment, Wheeler suggests that, like the word the player who left the room is looking for, the electron, in the experimental setup, is neither a particle nor a wave, *before* the physicist chooses to observe it. And, in that sense, it is unreal existing in a kind of indeterminate quantum limbo.

"Not until you start asking questions, do you get something," Wheeler said. "The situation cannot declare itself until you've asked your question. But the asking of the question precludes the asking of another. So if you ask where my great white hope presently lies—and I always find it interesting to ask people what's your great white hope—I'd say it's in the idea that the whole show can be reduced to something similar in a broad sense to this game of 20 questions.

Wheeler has condensed these ideas into a phrase that resembles a Zen koan: "the it from bit." In one of his free-form essays, Wheeler unpacked the phrase as follows: "...every it—every particle, every field of force, even the space-time continuum itself—derives its function, its meaning, its very existence entirely—even if in some contexts indirectly—from the apparatus-elicited answers to yes-or-no questions, binary choices, bits."[45]

In the 1960s, Wheeler was instrumental in popularizing the anthropic principle which states the universe must be as it is because, if it were otherwise, we probably would not be here to observe it. He was also one of the first prominent physicists to posit that reality might not be wholly physical which, for most physicists, prominent or not, is tantamount to heresy. He suggested, strongly, that, in some sense, the entire cosmos might be a *participatory* phenomenon, *requiring* the act of observation—and thus consciousness itself.

And here, once again, a loop reveals itself pointing unambiguously toward the dimensional structure of consciousness and the disquieting conclusion that—perceptual consciousness—like it or not, is wholly solipsistic.

Personally, when I read descriptions of reality by philosophers suggesting that it depends on observers leading inexorably to the abyss of solipsism, I find the arguments speculative and unconvincing. But, when a renowned physicist, of the stature of John Archibald Wheeler comes to the same conclusion as the philosophers, I find myself intellectually forced to reassess my opinion because his opinion is grounded in the results of disciplined experiments and the mathematics attached to the results of those experiments. Instead of saying "I believe reality might be a participatory phenomenon because after thinking about it for a long time that is the conclusion I've reached." He is saying "I believe reality might be a participatory phenomenon because *after thinking about the results of carefully controlled experiments conducted in the laboratory and the mathematics associated with those results I am led directly to that conclusion.*"

Physics is the hardest of the hard sciences and, for me anyway, it is the physics that forces me to reassess my opinion and essentially seals the deal.

Wheeler saw intriguing links between physics and information theory which the mathematician Claude Shannon invented in 1948.

Physics is built on an evanescent, elementary, indivisible entity—the quantum—which is *experimentally* defined by the act of observation.

Information theory, too, is built on an evanescent, elementary, indivisible entity—the binary unit or bit which is a message representing one of two choices, yes or no, heads or tails, one or zero.

Information is not physical. The binary choices heads or tails, yes or no, one or zero, all stand for the same thing, in terms of information. And, while they can be instantiated or reflected in the state of physical objects like a coin, a light on a display or a silicon gate in a computer's memory the information itself is never physical. That is what makes Wheeler's *it from bit* proposition so interesting. Looked at from another perspective Wheeler suggested that every *thing* comes from nothing. We set up double slit experiments with photons or electrons expecting them to behave like macroscopic objects behave but, instead, we discover they not only refuse to obey the laws of Newtonian mechanics, they refuse to reveal information about what they do or where they are until or unless we ask a yes or no question regarding what happened; and even then, what we discover is wholly dependent on what we choose to know about the event we are examining.

For hundreds of years, of all the quantum level phenomena we have explored, light—still—is one of the most interesting, vexing, baffling and enigmatic. It looks smooth but it is not smooth. What we are allowed to *know* about it; whether it is composed of discreet, individual bits or photons or sinusoidal waves is entirely dependent on how we set up our experimental apparatus and on the yes or no questions we ask about what we observe. The observations we make and the questions we ask wholly determine what the light will do and be.

Dreams

As multicellular organisms, our experience of the universe, while awake and active is the result of being embedded in perceptual consciousness and participating in the flow of observational consciousness. When we are awake, the self is fully engaged and in charge and we find ourselves busy with the tasks associated with *doing* as we rotate spatial dimensional axes orthogonally by doing things like walking from our living room to the kitchen or driving to the store.

When we communicate with each other by having a conversation, reading, using social media like Facebook and Twitter or watching television, we are engaged in the flow of observational consciousness. During all this activity the self or "I" is in the driver's seat.

As you read these words, your *self* is engaged in the flow of observational consciousness; trying to make sense of the ideas I am attempting to deal with and reveal.

But, what happens when we dream? When I dream the people places and things I encounter are every bit as real as when I am awake. In fact, when engaged in a dream, most people believe they are awake. The central difference between what we experience when awake and when we are in dream space is, the experience in dream space, is disjointed and the scenes, characters and surroundings change in ways that are confusing and unpredictable. We also often find ourselves in situations that, while awake, would be impossible or would not make any sense. But we proceed *as if* nothing out of the ordinary is going on. If the dream is particularly disturbing or frightening, when we awake, we realize it was *just a dream* and breathe a sigh of relief.

Prior to the work done by psychiatrists, John Allan Hobson and Robert McCarley in the 1970s, what we actually knew about sleep and dreams was severely limited and highly speculative. We had no scientific knowledge associated with the physiological aspects of sleeping and dreaming and we did not understand that there are specific and measurable neurophysiologic brain states associated with being awake and being asleep.

Our ideas about dreams were mythical and magical and even serious attempts at understanding and linking dreams to Freudian concepts regarding psychoanalytic psychiatry; while, at the time, *seemed* to make sense, were inaccurate and so highly speculative as to be scientifically and clinically useless. In fact, psychiatry practiced following Freudian principles of psychoanalysis has been scientifically discredited to the extent that, now, no more than three or four Harvard students apply for a residency in Freudian psychoanalysis.[46]

Human beings have had an intense interest in dreams, their content and analysis of their content for centuries and today this is still true. When I Googled the search term "Dream Interpretation Books" I got 1,460,000 hits. Clearly people are intensely interested in dreams and interpreting their meaning. But, if we hark back to Zion's axiom, we are reminded that, among other things, so many books being available on any subject is a clear indication that we know very little about the subject of those books. One reason is, until very recently, nobody had done any careful or rigorous scientific analysis of sleep or dreams. We assumed the sole purpose of sleep was rest. Many assumed that sleep was time wasted and that aside from its restorative value represented time which could have been better spent on more productive activity.

During the 1950s, researchers began to study sleep and dreaming. At the University of Chicago Eugene Aserinsky discovered that young children had periods of brain activation and rapid eye movements while they slept. And William Dement suggested that this REM *rapid eye movement* sleep was the psychological substrate of intense dreaming.[47]

We now know that rapid eye movement during sleep is a reliable indicator that the sleeper is dreaming. When we dream, our eyes move rapidly back and forth behind our closed eyelids.

Hobson was a practicing psychiatrist before he began to pursue the scientific aspects of sleep research. It made perfectly good sense to him to explore what was going on in the brains of animals and people while they were asleep and while they were dreaming.

His status as a physician and psychiatrist with deep experience treating patients with severe psychoses positioned him to be able to bring his unique experience to bear when analyzing dreams, their content and the neurophysiology associated with the dream state.

We spend about a third of our lives sleeping. Everybody dreams. Dreaming is so common that most of us don't pay much attention to it. But for Hobson and McCarley, paying attention meant much more than simply noticing what was happening. As psychiatrists and disciplined researchers, they brought their formidable talents and resources to the task of analyzing dreams both in terms of psychoanalytic principles and in terms of the neurophysiology associated with the dream state.

As a practicing psychiatrist with years of experience psychoanalyzing and treating patients with severe forms of neurosis and psychosis, for Hobson sleep research opened a window on the psychosis-promoting mechanisms of the normal brain. As he proceeded with his sleep research the formal approach he used revealed the following specific properties definitive of dreaming.

- "Hallucinations, especially of vision and movement but also hallucinations involving other sensory systems. Without any sensory input, we see things clearly that are not there.
- Delusions. When we dream, we harbor the conviction that we are awake. We are also prone to uncritical belief in physically impossible and/or improbable dream scenarios. During dreams, we almost never correct these errors of judgment.
- Cognitive abnormalities. These include loss of declarative memory (amnesia within the dream), loss of orientational stability (the discontinuity and incongruity of dream bizarreness), and marked reductions of declarative memory of the dream upon awakening (amnesia). Thus, when dreaming we are as cognitively impaired as we are perceptually enhanced. This is exactly true of patients with so-called mental illness.
- Emotional intensifications. Dream elation, anger, and anxiety are three emotions which are often perturbed in schizophrenia, in major affective disorders, and in neurotic conditions."[48]

Everyone knows their dreams are bizarre. For most of us, that's just the way it is; we may take note of the strangeness associated with our dreams but we rarely pay attention to or think about that strangeness. When dreaming we find ourselves in places and situations that would make no sense at all, if we were awake. We interact with people we know, think we know and don't know and we experience intense emotions.

In thinking about all this, one question that arises is *what does it mean?* Some people, including Sigmund Freud, attempted to attach meanings to dreams and interpret them in ways that reflected what was going on in the normal waking state of the dreamer. This kind of dream interpretation, whether done by Freudian psychoanalysts or gypsies or witch doctors, has been thoroughly discredited as having any substantial or reliable scientific validity. It is all mumbo-jumbo and hocus-pocus. But, looked at through the lens of a dedicated researcher with deep experience treating severely disturbed patients, different pictures and answers begin to emerge. From one perspective dreams are simply what humans and animals do when they enter the REM stage of sleep. Hobson, however, saw things differently. His scientific conclusion regarding REM sleep was that it should be viewed as a normal condition of the brain. But, his training and experience as a psychiatrist also led him to the conclusion that while REM sleep is certainly a normal condition of the brain, it also serves as the substrate of the cardinal features of psychosis that characterize many of the most severe forms of mental illness. And his conclusion regarding these observations was:

> The fact that we are all capable of dreaming means that we are all capable of a completely normal and natural form of madness.[49]

Most of us have had the experience of watching someone being portrayed in a movie as being completely and irreversibly insane and asking ourselves what it must be like to be in such a mental state. Some of us have experienced direct interaction with people, sometimes loved ones, who are certifiably insane and asked ourselves the same question. Hobson's startling and, frankly, somewhat disturbing conclusion is, if you truly want to understand madness all you need to do is pay careful attention to what goes on when you are dreaming. Our dreams are characterized by hallucinations, delusions, bizarre cognition, amnesia and confabulation.

Every night, when our brain state changes from its normal waking state to REM sleep, we become quite floridly psychotic. Put differently, every time we dream we go nuts.

Hobson's observations regarding dreaming have deep implications for psychiatry and helping to expand our understanding of the specific differences between normal and abnormal brain states.

Madness is usually viewed as a structural aberration of normal brain function but Hobson's conclusions regarding what happens to our brains when we dream shows that madness can be functional as well as structural and that it need not *always* be an indication that we have abnormal genes, biased neuronal receptors or suffer from traumatic life events.

It is certainly true that when we dream we become certifiably insane, but that does not mean that dreaming is abnormal. And, conversely, madness in individuals who are awake is certainly not normal. What Hobson observed as important about all this is dreams and awake state madness share *formal* properties and that understanding how and why the brain dreams during sleep not only can but will provide a useful framework for discovering how and why the brain becomes psychotic in the awake state.

Since I am not a psychiatrist, it would never have occurred to me to associate dreaming with insanity. But now that I understand the connection I must admit that I find it more than just a little bit disturbing. Now, when I kiss my wife good night, I think to myself, *good night sweetheart, I'm going to go to sleep and dream now and become floridly insane for awhile but when I wake up I'll be perfectly normal.*

I have always been interested in and curious about mind and consciousness. Thousands of books have been written on the subjects and countless philosophers have argued and continue to argue about both subjects. Here's Hobson's take on the subject of mind:

> ...consciousness must be our subjective awareness of an activation state of brain neurons. This conclusion followed inevitably from the obvious fact that visual perception was nothing more or less than activation states of certain brain cells. Light edges, bars, and shadows do not pass beyond the retina. From the end organ on in—and out again to the movements of our body—it is all nothing more or less than neural information that is processed. Of course we need to know more about how complex images are built up from the bits and pieces of visual code. We need to know more about how such images are perceived and then tied to memories, to feelings and to thoughts as we respond to real objects or create inner visions. And we need to know how these brain processes are translated into motor action. But once we recognize the radical reality of neural representation, the

> mind-brain problem vanishes. It is a false problem. The mind is an activation state of the brain.[50]

In thinking about that quote, I was reminded that Hobson's Cartesian imprinting is as strong as the imprint that most of us have. The sentence that reveals the imprint is "*We need to know more about how such images are perceived and then tied to memories, to feelings and to thoughts as we respond to* real objects *or create inner visions.*"

Clearly Hobson is not a solipsist. He believes objects are real and exist "out there" and that somehow our brain state allows us to perceive and interact with those objects. Physics, quantum mechanics and the dimensional structure of consciousness, paint a different picture of reality, which I have talked about earlier.

Another thing that Hobson's quote reminded me of was the tiny shiny simms and simmballs crashing and bashing about on Douglas Hofstadter's Careenium.

Our brain states and levels of awareness associated with those states are dependent on what area or areas of our brains are involved and on the chemical neurotransmitters which carry signals between the neurons responsible for a given state.

When we are awake, and what we recognize as our *self* arises in order to do, our prefrontal cortex is active and involved and the neurotransmitters doing the work are predominantly aminergic, which means mediated by the chemical molecules histamine, norepinephrine and serotonin.

When we dream, our prefrontal cortex deactivates and cells toward the back of our brain including the pons and brain stem become active. Our brain state changes both in terms of where the cells are located and in terms of the neurotransmitters responsible for the activity among and between the neurons that are firing. Where, in our awake state, the neurotransmitters are aminergic, when we are asleep and dreaming, the primary neurotransmitter is acetylcholine, which is cholinergic rather than aminergic. This discovery by Hobson and McCarley explains why our thinking is so disjointed and meager while we are dreaming.

"The poverty of dream thought is our subjective awareness of deactivation of the higher brain functions involved in directed thought.[51]"

Our *executive* prefrontal brain state is no longer available and, even though we are aware of that, we proceed, in the dream, as if everything is all right. That explains why dreams are so bizarre. The primary neurotransmitter enabling the neurons responsible for what's happening in our dream to fire is acetylcholine. While we're in this *cholinergic* brain state we no longer have access to the executive functions associated with our prefrontal cortex because it is, in a sense, asleep. Compared to our normal waking state we are, for all intents and purposes, insane. We are having exactly the same kinds of experiences that people afflicted with sever psychotic conditions like schizophrenia, severe depression and bi-polar disorder have when they are fully awake.

Some people are able to become aware of the fact that they are dreaming. When that happens, the prefrontal cortex becomes active (awakens?) and the dream becomes lucid. I have had this experience several times and it is truly unusual. I get to take control of what's happening in the dream but, if I allow myself to become fully engaged, I immediately lose lucidity and return to a normal that is, *floridly insane* brain state which is fully cholinergic.

In an email, Dr. Hobson agreed with my conclusion that when a dreamer becomes lucid, the brain activity which enables lucidity is aminergic. Prefrontal cortex activity ensues and the neurotransmitters norepinephrine, histamine and serotonin enable enough executive brain function to allow me to recognize that I am dreaming and take control of what's happening. But, if I allow myself to become engaged, in the same sense that I'm engaged while awake, I lose lucidity and return to my previous cholinergic brain state mediated primarily by acetylcholine.

As Hobson and McCarley studied what was going on in the brain in the dream state, they focused on describing and measuring the nature of the conscious experience that dreaming represents rather than attempting to interpret dream content. This was a radical change to dream psychology and they suggested the idea of brain-mind isomorphism.

For most people, the term isomorphism makes their eyes glaze over. It's really not that difficult to understand. Objects can be isomorphic. A ring is a good example. My wedding ring is made of titanium. If I had one made of wood, while it would certainly look different it would be isomorphic with the one made of titanium. House cats, lions, cheetahs, tigers and panthers are isomorphic. Hobson and McCarley are suggesting that the same kind of isomorphism exists between mind and brain and that it is therefore possible to map mind to brain and back from brain to mind. The reason they give for this isomorphism is all mental events are subjective experiences of brain events; *neurons firing*. If that's true then, at some level, there must be a demonstrable similarity of form or *isomorphism* which links the two domains of discourse.

They didn't stop there but went even further by asserting that the two domains of discourse differed only at the relatively superficial level of language description. And their conclusion was:

> "at a deeper level, brain activation was mind activation. In other words, our theory aspired to a fundamental monism. There is only one system, and it is fully integrated. I call it the Brain-Mind."[52]

Philosophers would argue and have argued endlessly about this suggestion of monism and their arguments, while interesting, have a tendency to cancel themselves out. Hobson offers three statements in an attempt to clarify what he means by Brain-Mind isomorphism:

- Dream vision is our subjective awareness of visual brain activation.
- Dream emotion is our subjective awareness of emotional brain activation
- The poverty of dream thought is our subjective awareness of deactivation of the higher brain functions involved in directed thought.[53]

Hobson and McCarley avoided complicated psychological theories in an attempt to explain dream vision, dream emotion and the poverty of dream thought.

Instead they focused intensely on what the brain was actually doing during REM sleep which is when our most intense dreams occur.

This formal approach to studying dreams led them to conclude that:

- Dream form is not in any way causally connected with what happened to us yesterday
- Dream form is not in any way causally connected with what happened during childhood
- Dream form is not in any way causally connected with what might happen to us in the future

Before Hobson and McCarley carried out their careful and rigorous investigations, most of what we knew about dreams was speculative and interpretive based on the content of those dreams. Even Sigmund Freud tried to attach meaning to the content of dreams and then to map that meaning back onto psychoanalytic principles associated with the ego, the superego and the id. Now we know it was all hogwash.

Instead of allowing themselves to be lured into the trap of focusing on dream content, and becoming entangled in content issues related to what happened to the dreamer as a child, what happened yesterday or what might happen in the future, they focused instead on dream form and reached a much less squishy and more robust conclusion regarding what actually happens when we dream. They looked at what areas of the brain are active when we dream and identified what neurotransmitters were involved. They also identified what areas of the brain are inactive when we dream and discovered that the aminergic neurotransmitters responsible for awake state cognition are not responsible for what we experience while we are dreaming. This is science in the grand tradition.

In scientific terms, Hobson's and McCarley's work suggests replacing Freud's Disguise-Censorship Model of dreaming with their Activation-Synthesis hypothesis which, because it is rooted in hard science, it, by definition, renders the Disguise-Censorship model obsolete.

In summary Hobson concluded the following:

> In 1975, McCarley and I began to pursue dream psychology in earnest, even if we were neither paid nor praised for it. Our 1983 discovery that dream bizarreness could be reduced to orientational instability led us to the Dreaming as Delirium concept. Delirium is the pathological state that many of my patients at Bellevue had evinced. Delirium consists of visual hallucinations, disorientation to time and place, recent memory loss, and confabulation. Dreaming, which is normal, shows all four of these signs of and is thus delirium by definition. In REM sleep, we all become delirious, because the brain changes its own chemistry as radically as it ever does under the influence of the drugs or alcohol that cause pathological delirium.[54]

One question that arises when we consider the Hobson/McCarley Activation-Synthesis hypothesis of dreaming and the brain states associated with it is, if it is correct, what evolutionary purpose does it serve? How does it help us to survive in the same sense that being able to see, hear, smell, taste and feel help us to survive?

Hobson's answer is it's possible that the function REM sleep serves is to prepare the brain for waking consciousness. And that, even while we are awake, it acts as a virtual-reality generator. But, it is only when we dream that we get to directly experience this creative process in all its glory.

How does this square with our explorations of the dimensional structure of consciousness? When I hear the squirrel chittering in my backyard and go to the window, I see him there, on the lawn. If I heard him in a dream and went to the window, I might see a squirrel but it's just as likely that I might see an elephant, giraffe, or dinosaur because, during a dream, I'm watching the virtual-reality generator do its work without being aware of what mental state I'm in. My prefrontal cortex is not active and the neurons firing in the back of my brain, creating the chittering sound, are being enabled by acetylcholine. Without access to the aminergic neurotransmitters histamine, norepinephrine and serotonin I lack the executive brain function which allows me to become and remain aware of my state of awareness. Put differently, I don't know that I'm dreaming and assume I'm awake but the part of my brain that, while I'm awake, lets me *know* that I'm awake, isn't available to me in the dream.

Hobson is suggesting that, even when we are awake, this virtual-reality generator is still working but, with the help of our prefrontal cortex and the availability of aminergic neurotransmitters we're able to stitch our experiences together in a way that is more stable than what happens when we dream. Our prefrontal cortex is fully engaged using the aminergic neurotransmitters histamine, norepinephrine and serotonin. Our *self* has arisen and is active and we go about our business unaware that, in the background, the virtual reality generator Hobson identified is enabling the experiences we consider normal.

This model of brain function points strongly toward our earlier explorations of the dimensional structure of consciousness. To me it suggests that the *out there* we assume exists in our awake brain state is the same *out there* we experience when we dream and, if that's right it's another strong indicator that there is no *out there.*

The cardinal difference between the *out there* we experience when awake and the *out there* we experience when we dream is our brain state. Prefrontal cortex aminergic brain activity (the self?) provides the executive function necessary to damp down the craziness of the virtual reality generator which is essentially *always* engaged and functioning. If this is true, then clearly dreaming provides a survival advantage even more important to our survival than our ability to see, hear, feel, smell and taste.

Hobson believes that, absent dream consciousness, we cannot have waking consciousness. And while he admits this hypothesis is speculative, the results of the careful and scientifically disciplined research he and McCarley carried out indicate that his hypothesis is well on its way to becoming theory; it has legs. I believe it also fits quite nicely with and supports Samuel Avery's dimensional structure of consciousness. So now, in addition to the physics and quantum mechanical principles which support the dimensional structure of consciousness we have the neurophysiologic and psychological underpinnings revealed by the research of Hobson and McCarley.

Thus far, in this chapter, we have explored consciousness, dreams and the awake state in adults, as revealed by the work of Hobson and McCarley. But, on the subject of consciousness, Hobson offers yet another conceptual hypothesis he calls protoconsciousness. And by 'proto' he means both prior to and foundational of consciousness as we normally think of the word and what it means.[55]

He talks about his own gestation in his mother's womb and suggests we begin our thought experiment by asking that we first subtract the content of what we have good reason to believe derives from awake-state consciousness and see what might be left over. So, after taking away all the things we normally associate with being awake and conscious he suggests that first and foremost is some primordial sense of self. "My brain is activated, therefore I am"?[56] Shades of Descartes.

To the question, of whether we are sentient *in utero* he opines that the likelihood of any sentience worthy of the name is highly improbable. Instead he offers that when his brain became activated he moved. By the third trimester of his mother's pregnancy he was capable of a wide range of movements such as moving his arms and legs, sucking his thumb and eye movement that, at times, occurred in REM like clusters.

Whether or not the movements he made while floating in his mother's womb were voluntary or simply reflexes is a moot question but with the non invasive imaging technology we now have available we can find out. The point he tries to make in describing his *in-utero* existence is that, as he developed, he took responsibility for actions like kicks, hand waving and thumb sucking that, at first, may have been purely reflexive. And, through this process of taking responsibility an "I" or "self" began to emerge. His brain was in a state of pure protoconsciousness.

After he was born, his brain (the primordial "I" or "self") that developed along with the rest of his body was prepared to move gradually from a state of protoconsciousness to one of waking consciousness and dream consciousness. He cried in protest at having been ejected from his previously warm and comfortable place of existence.

> The dawn of consciousness must be gradual. By the time I was a year old, my mother was convinced that I was conscious (when awake) even if she never dreamed that I was dreaming when I was asleep. David Foulkes asserts that adult type dreaming begins between the ages of 6 and 8 years. I think it may begin between 5 and 7, but that is still quite late. My point here is that the individual evolution of awake-state and dream consciousness is a long, gradual process that cannot be precisely demarcated. In this sense, the origin of consciousness is analogous to the origin of species, which has a long ill-defined history.[57]

Hobson believes that by the third trimester of pregnancy a fully developed protoconsciousness platform emerges in the fetus. After being born, the protoconsciousness which developed *in utero* is dampened. As more and more time is spent in the awake state, the dampening process proceeds. The interaction of the residuals of protoconsciousness (dreaming and primary consciousness), gradually give way to a progressive and fully enriched waking (or secondary) consciousness, and for at least five years a parallel development of dreaming consciousness. He also believes there is lifelong communication of content issues between awake state consciousness and dreaming consciousness, the interactional dynamics of which are relatively fixed between the ages of 15 and 60. After 60, both systems gradually deteriorate.

"To sleep perchance to dream—ay, there's the rub." Even in the time of Shakespeare, that line from Hamlet's soliloquy probes the connection between sleep and dreaming. In the end, if Hobson is right, our dreaming brain may be the virtual reality generator that we ultimately learn to dampen when we move from protoconsciousness to fully enriched waking or *secondary* consciousness. And then, after an appropriate amount of time has passed, we learn to dream.

Our dreams *reveal* the virtual reality generator allowing us to witness what it does but an interesting question about that becomes: While we are dreaming, if it is not a lucid dream, who is participating in what is going on in the dream?

To me, it *feels* like the same *self* or "I" that arises during my normal awake state. When, occasionally, I become lucid during a dream, my prefrontal cortex becomes activated (awakens?) and I get to recognize that I am dreaming and can assert some control over the situation but, if I allow myself to become fully engaged, I lose lucidity. My sense is there is something important about that, but I don't know what that something is. My guess is it is somehow related to meditation and how difficult it is, regardless of what method you use for meditation, to dampen the *self*, strip away the labels and *see* the chaos.

Perhaps dreaming is the neurophysiologic mechanism we use to strain the chaos, turning it into a substrate for waking consciousness. If true, it fits with Hobson's belief that without dream consciousness, we cannot have awake-state consciousness.

Another visit to quantum reality

We can see, smell, feel, taste and touch objects that exist on the macroscopic level of reality where we exist. For objects that exist in the cosmos, like planets, stars, galaxies, pulsars, black holes and neutron stars, we can see some of them, infer their existence by observing certain physical phenomena and even directly feel them like when we feel the warmth of the sun on a summer's day.

The common denominator on the macroscopic and cosmic levels of reality is space–time. Space–time arises and the macroscopic universe unfolds in the touch of light.

But, on the quantum level of reality, space and time disappear; both in terms of the results we get when we perform experiments with quantum level objects like photons and sub atomic particles and in terms of the equations we use to explain what we observe when we perform those experiments.

Physicists have proven that atoms, subatomic particles and photons exist, on the quantum level and that the quantum level of reality underlies or is the substrate of both the macroscopic or classical level of reality where we spend most of our time and the cosmic level of reality which includes deep space, the planets, galaxies and all the other astronomical entities astronomers and astrophysicists have discovered.

On the macroscopic or classical level of reality, for the most part, our experiences are relatively stable. When I reach for the cup I see sitting on my desk I can feel it when I touch it. When I hear the squirrel chittering in my backyard, if I go to the window, I can see him and, sometimes, I can even get him to come to my outstretched hand to grab some tasty peanuts. For most of us, most of the time, these kinds of common everyday experiences tend to reinforce our understanding of how the world works and they support the myth of materialism handed to us by Descartes. We use these experiences to frame and explain what's going on.

But, when we begin to carefully examine our macroscopic experiences the stability we've come to rely on begins to crumble because we discover we cannot explain phenomena that, on the surface appear not to need any explanation; like why do objects have mass? And, how are massive objects able to resist acceleration even if we are trying to move them in outer space where they are, by definition, weightless?

These kinds of phenomena present a direct challenge to our Cartesian understanding of what's going on and undermine the myths we rely on to help us understand what is *really* going on. They raise questions about what we are observing; and when we try to use common sense to explain these phenomena the answers we get are enigmatic and unsatisfying.

Part of the reason this is so is when we do experiments with quantum level objects like photons or subatomic particles, we cannot see, feel, smell, taste or touch them. We must use other quantum level entities to disturb the particle we are trying to observe. And, when we get an observable result what we *see* is indirect; the particle's spin changes or its polarization reverses. We are not allowed to know everything about the particle. If we know its position we are denied knowledge about its momentum and vice-versa. So we must infer what our observations tell us and, like it or not, accept the uncertainty attached to such endeavors.

As I suggested earlier, the reason this is so is when we perform experiments using photons or subatomic particles, we are attempting to probe what is happening on the quantum level of reality and the results we get provide an indirect glimpse of the chaos. The results are enigmatic and counterintuitive because we are watching phenomena unfold on the quantum level using macroscopic instruments.

When we entangle two subatomic particles, we find that regardless of how far they are separated in space, if we disturb one entangled particle the other particle reacts instantaneously. Classical physics and special relativity forbid instantaneous reactions between entangled particles separated in space yet countless experiments have proven that the reactions are indeed instantaneous and would be even if the particles were at opposite ends of the cosmos.

The entangled particles are operating on principles associated with the quantum level of reality where they exist. And, as we have seen, space and time do not exist on the quantum level. The experimental results we get with entangled photons and subatomic particles are enigmatic and counterintuitive because, once again, we are indirectly glimpsing the chaos when we perform such experiments.

The quantum level of reality in which neither space nor time exists is one level above the void where neither space, time nor subatomic particles exist. In the void, subatomic particles pop in and out of existence from what physicists refer to as the quantum foam.

It is extremely difficult to try to form a mental picture of the void because, the only tools we have to form mental pictures are our experiences in the classical or mid level of reality which includes space and time. Some people think of space without any galaxies, planets or stars seeing it as just a vast, endless black emptiness but the problem with that is a vast endless black emptiness implies space and the void has no space. The void is the absence of *everything*. It is the chaos.

On the quantum level objects like subatomic particles and photons exist that we can detect and manipulate. When we began doing that some of the results we got were so counterintuitive that the early pioneers of quantum mechanics created thought experiments to attempt to understand what they were observing and to demonstrate that, what they saw made absolutely no sense when filtered through our understanding of how the world works in the mid level or classical level of reality.

One of the most famous and best known of such thought experiments was formulated by one of the early pioneers of quantum mechanics, Erwin Schrödinger.

It was Schrödinger who, in 1935 coined the quintessential quantum effect entanglement. His equations showed that, once entangled, two subatomic particles or photons behave as if they were one entity and the implications of entanglement were, at the time, so bizarre and counterintuitive that, when Albert Einstein examined Schrödinger's work, while he could find no fault with the mathematics, he referred to entanglement as *spooky action at a distance*.

In an effort to demonstrate how strange the implications of entanglement were Schrödinger came up with the now famous thought experiment known as Schrödinger's cat.

After examining and thinking about his equations that described entanglement he realized that it applied to macroscopic objects as well as subatomic particles and photons. His thinking went something like this: if an atom of a radioactive element like radium could exist in a state of superposition of being both decayed and undecayed simultaneously, then a macroscopic object like a cat placed in the same environment as the atom of radium would become entangled with that atom and also enter a state of superposition.

In the thought experiment Schrödinger imagined a box, large enough to easily accommodate a cat. He also imagined a device, in the box, which had a small bottle of cat poison which, if broken, would kill the cat and a mechanism which would be triggered if the radium atom decayed. The decay of the radium atom would cause the mechanism to break the bottle and kill the cat.

Here's how the thought experiment works. If you take a live cat and place it in the box and close the lid on the box, the cat immediately enters a state of superposition of being both alive and dead because its fate is entangled with whether or not the radium atom decays. We now have two entities in superposition, in the box; the macroscopic cat and the quantum level radium atom.

Once the lid is closed, in order to find out whether the cat is alive or dead we must open the box to *observe* its condition. Opening the lid of the box collapses the quantum probability wave associated with the cat being either alive or dead but until or unless we do that the cat remains in a state of superposition of being both alive and dead, simultaneously.

Under normal circumstances we never encounter cats being both alive and dead simultaneously. They are either one or the other but never both.

The point of Schrödinger's morbid thought experiment was to show how strange the concepts of superposition and entanglement are when applied to a macroscopic object like a cat placed in a box which forced it into a state of superposition tied directly to the radium atom, tied to the mechanism which would break the bottle of cat poison if the radium atom decayed. Schrödinger's equations showed, unequivocally, that the only way to discover whether the cat was dead or alive after the lid was closed was to open the lid and make an observation. But, until that observation was made the cat remained in a superposed state of being simultaneously both alive and dead.

Schrödinger's cat thought experiment was meant to show that while the equations demonstrated that entanglement and superposition were phenomena that applied to and could be observed when dealing with quantum level entities, it made little or no sense to assume they also applied to macroscopic objects like cats. But it's important to keep in mind that was in 1935.

In the middle or classical level of reality, where we spend most of our time, the reason we never see cats being both alive and dead simultaneously or our coffee cup being both on the desk and in the cupboard simultaneously is because the complex and layered interactions macroscopic objects have with their surroundings hide whatever quantum effects may be going on. Quantum information regarding whether the cat is alive or dead or the location of our coffee cup rapidly leaks into the environment in the form of photons and the exchange of heat. That's why we don't see quantum level information associated with macroscopic objects.

Characteristic quantum level phenomena entail combinations of different classical states like the coffee cup being both on the desk and in the cupboard and the cat being both alive and dead. These combinations are what tend to dissipate in the mid level of reality. In physics-speak this information leakage is the essence of a process called decoherence.

Decoherence is what causes the probability wave to collapse forcing an object that is in a state of superposition—out of superposition. When we open the lid of the box the cat is no longer in a state of being both alive and dead simultaneously. Our observation collapses the probability wave that sent the cat into a superposed state because we now *know* whether it is alive or dead. The decoherence reveals which of the two states the cat is in.

Schrödinger was relatively certain that quantum level phenomena like superposition and entanglement didn't apply to macroscopic objects like cats and coffee cups and his thought experiment was meant to show that, even though his equations suggested strongly that superposition and entanglement did apply to macroscopic objects, assuming that it truly did, didn't make much sense.

Fast forward now to the present and we find, once again, that physicists, faced with the intellectual itch caused by all the weirdness associated with quantum mechanics including superposition, entanglement and subatomic particles behaving as if they were waves, as photons do when we perform double slit experiments; we discover that the physicists contrived experiments to help scratch that itch and, as happened earlier, their results revealed that their assumptions about quantum level phenomena applying only to subatomic particles at very low temperatures were wrong.

What they discovered was: "Quantum effects are not limited to subatomic particles. They also show up in experiments on larger and warmer systems."[58] Here are some examples.

In 1999, Markus Arndt and Anton Zellinger et al. at the University of Vienna, "[o]bserved interference pattern for buckyballs, showing for the first time that molecules, like elementary particles, behave like waves at a temperature of 900–1000 kelvins."[59]

This was a variation on the experiment Louis De Broglie did with electrons where he demonstrated that they behaved just like photons when subjected to the double slit experimental setup. The key difference here is we're talking about objects that are orders of magnitude larger than electrons. Buckyballs are very large molecules.

In 2009, Alexandre Martins de Souza et al. at the Brazilian Center for Physics Research "[d]educed entanglement of trillions of atoms (or more) from the magnetic susceptibility of metal carboxylates at a temperature of 630 K"[60]

In 2010, Elisabetta Collini et al. at the University of Toronto, University of New South Wales and University of Padua, "[f]ound that quantum effects enhance photosynthetic efficiency in two species of marine algae at a temperature of 294 K."[61]

In 2011, Stefan Gerlich and Sandra Elbenberger et al. at the University of Vienna "Set a new world record for observing quantum effects in giant molecules, including an octopus-shaped one with 430 atoms at a temperature of 240–280 K."[62]

In 2010, Leonardo DiCarlo and Robert J. Schoelkopf et al. at Yale University and the University of Waterloo, "[e]ntangled three quantum bits in a superconducting circuit. The procedure can create quantum systems of any size, at a temperature of 0.1 K."[63]

In 2010, Aaron O'Connell and Max Hofheinz et al. at the University of California Santa Barbara, "[c]oaxed a tiny springboard about 40 microns long (just visible to the unaided eye) to vibrate at two different frequencies at once at a temperature of 25 millikelvins."[64]

In 2005, Hartmut Haffner and Rainer Blatt et al. at the University of Innsbruck "[e]ntangled strings of eight calcium ions held in an ion trap. Today the researchers can manage 14 at a temperature of 0.1 mk."[65]

In 2009, John D. Jost and David J. Wineland et al. at the National Institute of Standards and Technology "[e]ntangled the vibrational motion—rather than the internal properties such as spin—of beryllium and magnesium ions at a temperature of 0.1 mk."[66]

As discussed earlier, physicists are interested in observing what is. They don't like dealing with nonobjective phenomena like mind, consciousness and qualia and if they find themselves in a position where experimental results suggest that nonobjective phenomena, like the presence of an observer, have a direct and measurable impact on the outcome of their experiments, they become uncomfortable and apologetic.

They would very much like to stick to the belief that physics is physics. It describes what happens in the *real* world. And things that are clearly not physics like mind, consciousness, qualia and observers simply get in the way of the purity they've always assumed was attached to what they do.

But, in the end, it's their cumulative curiosity—that nagging intellectual itch—that gets the better of them and forces them to probe what their equations are telling them with *objective* experiments. When those experiments demonstrate that what their equations suggested no matter how unintuitive or bizarre are physically observed under carefully controlled laboratory conditions they find themselves between the proverbial intellectual rock and hard place.

All this makes me both nod and smile because it reminds me of the quote from *Transcendence of the Western Mind*, "nobody will be more surprised than the physicists."

The weirdness associated with quantum mechanics has pervaded physics since the early 20th century. And, in a very real sense that weirdness is always with us even though we don't see or notice it. The situation is similar to the one associated with special relativity and the way time is dilated when we begin to accelerate at speeds approaching the speed of light. For the person accelerating, time slows down relative to someone who is not accelerating as we have proven with atomic clocks on airplanes and with subatomic particles accelerated to speeds approaching c.

But it is important to keep in mind that the time dilation associated with acceleration is always with us, even as we drive our cars past each other or walk past one another on the sidewalk. It is so small that we do not notice it but it is, nonetheless, always there.

These recent explorations of the implications associated with living in a quantum world are forcing physicists to recalibrate their thinking about concepts that, not so long ago, they believed they understood very well; concepts like space, time, gravity, and the existence of matter.

> Thus, the fact that quantum mechanics applies on all scales forces us to confront the theory's deepest mysteries. We cannot simply write them off as mere details that matter only on the very smallest scales. For instance, space and time are two of the most fundamental classical concepts, but according to quantum mechanics they are secondary. The entanglements are primary. They interconnect quantum systems without reference to space and time. If there were a dividing line between the quantum and the classical worlds, we could use the space and time of the classical world to provide a framework for describing quantum processes. But without such a dividing line—and indeed, without a truly classical world—we lose this

> framework. *We must explain space and time as somehow emerging from fundamentally spaceless and timeless physics* [italics mine].
>
> That insight, in turn may help us reconcile quantum physics with that other great pillar of physics, Einstein's general theory of relativity, which describes the force of gravity in terms of the geometry of spacetime. General relativity assumes that objects have well-defined positions and never reside in more than one place at the same time—in direct contradiction with quantum physics. Many physicists, such as Stephen Hawking of the University of Cambridge, think that relativity theory must give way to a deeper theory in which space and time do not exist. Classical spacetime emerges out of quantum entanglements through the process of decoherence.
>
> An even more interesting possibility is that gravity is not a force in its own right but the residual noise emerging from the quantum fuzziness of the other forces in the universe. This idea of "induced gravity" goes back to the nuclear physicist and Soviet dissident Andrei Sakharov in the 1960s. if true, it would not only demote gravity from the status of a fundamental force but also suggest that the efforts to "quantize" gravity are misguided. Gravity may not even exist at the quantum level.
>
> The implications of macroscopic objects such as us being in quantum limbo is mind-blowing enough that we physicists are still in an entangled state of confusion and wonderment.[67]

And so it goes. Nobody is more surprised than the physicists.

For me, the line, in the above quote, that stands out in satisfying relief from the physics and weirdness associated with quantum mechanics is the one I italicized: *We must explain space and time as somehow emerging from fundamentally spaceless and timeless physics.* When I read it I smiled and nodded because it says, although less eloquently, what Samuel Avery said, "the universe [including spacetime] unfolds in the touch of light."

Photons, which are spaceless and timeless quantum level entities, touch our retinal cells and spacetime arises. Our retinal cells don't *see* anything. They feel the touch of the photons and report being touched to our brains. We then *see* the photons and watch the photon screen unfold and spacetime arise.

The abject weirdness associated with quantum mechanics is now strongly suggesting to physicists that, like it or not, they must reexamine their dearly held fundamental concepts regarding the very existence of space and time. Albert Einstein described the spacetime continuum as a kind of fabric composed of both space and time. And, while on one level, associated with the mathematics and equations which describe spacetime in special relativity, it makes some semblance of sense to physicists; for most of us mere mortals, attempting to understand what the *fabric* of spacetime is or looks like proves to be a difficult if not impossible task.

The physics and equations tell us one thing while our macroscopic experiences conflict with the physicists' explanations.

We see emptiness between the objects that surround us and, for lack of a better word, we call that emptiness space. We look up into the cosmos and see emptiness between stars and galaxies and, for lack of a better word, we call that emptiness outer space or deep space.

We witness phenomena like aging and growth that strongly suggest time exists so, we assume it does and build devices we call clocks which are just representations of the earth rotating on its axis. We refer to what clocks depict as time and most people believe that clocks actually do measure time. But, when we focus our attention on time, in the same way physicists focus their attention on phenomena like heat, sound, photons and subatomic particles we find that we cannot detect time. When we look for it, it goes poof.

The more carefully we examine space and time the less reliable our assumptions about what they *really* are become.

We have grasped the physics and in so doing were led down the rabbit hole of quantum mechanics. And now, like Alice, we are trying to make sense of what we see and what we once thought we understood. The circle is closed with the physics included and intact. But the place we find ourselves in requires new and different kinds of explanations.

We're in the same place Copernicus was when he realized that the earth wasn't at the center of the universe and that it was the earth that was moving rather than the sun and stars. We're able to *see* that now because we believe it to be true. We've ventured out into the cosmos and walked on the surface of the moon. While there, we looked up and saw the earth rise. And, while everything looks the same to us now as it did to those who lived in the time of Copernicus, we *know* the truth about the relationship of the earth to the stars.

Copernicus made a guess and he was right. That guess shattered the geocentric myth which held sway, even long after he died.

The dimensional structure of consciousness is the same kind of guess. It files in the face of the materialistic dualism handed to us by Descartes.

We cling to Descartes' dualism for the same reason(s) we clung to the geocentric myth. It's comfortable and, for the most part, explains our experiences in the classical or mid level of reality. But, we've gone down the quantum mechanics rabbit hole and, suddenly, things no longer look the same.

We can choose to ignore what we've discovered but if we do, it will be at our own peril. There is no *going back*. The dimensional structure of consciousness provides a path for us to explore what we have discovered after going down the quantum mechanics rabbit hole.

We need to strap on our helmets, tighten our seatbelts and prepare for a bumpy, exciting and astonishing ride.

What about God?

God is a contentious subject because for every individual observer, God's existence or nonexistence is tied inextricably to belief.

Judeo-Christian and Middle Eastern concepts of God tend to be anthropomorphic and male. Roman Catholics believe in a tripartite God consisting of God the Father, God the Son and God the Holy Ghost. For them God is a man who created the universe and humans in his own image. Jesus was his son who became incarnate and died for our sins.

The Islamic word for God is "Allah," and like the Roman Catholic conception Allah is anthropomorphic and male.

Eastern religious traditions and practices like Buddhism, Confucianism and Taoism tend not to depend on anthropomorphic conceptions of God.

Eastern culture creates a mindset that is distinctly different from the one created by Western culture.

Where Western culture embraces Cartesian dualism, the existence of consciousness in individuals and the myth of the existence of matter, Eastern culture is much less heavily invested in such concepts.

Those who follow the Tao do not make reference to God in the same way(s) Westerners do. For a Taoist, the Tao simply is and Taoists understand that attempts to describe the Tao must always fall far short of what it truly is. The first verse of *The Tao Te Ching*, written 2,500 years ago by Lao Tzu, demonstrates what I mean.

> The Tao that can be told
> is not the universal Tao.
> The name that can be named
> is not the universal name.
>
> In the infancy of the universe,

there were no names.
Naming fragments the mysteries of life
into ten thousand things and their manifestations.

Yet mysteries and manifestations
spring from the same source:
The Great Integrity
which is the mystery within manifestation,
the manifestation within mystery,
the naming of the unnamed,
and the un-naming of the named.

When these interpenetrations
are in full attendance,
we will pass the gates of naming notions
in our journey toward transcendence.[68]

Lao Tzu understood that naming or attaching a label to his conception of the Tao could not, under any circumstances, represent what the Tao truly is. In the third stanza of the verse he uses a term which I interpret as being similar to what Westerners understand as God—The Great Integrity. But it also seems clear to me that The Great Integrity is synonymous with the Tao. Another thing that seems clear is Taoists do not envision or anthropomorphize the Tao with a person, male or female. Instead they *see* the Tao as something other than everything and anything else that can be perceived under normal circumstances.

It is possible to read the entire Tao Te Ching in less than an hour. I have read many different translations many times and discovered that it has an interesting quality in that, each time I read it, I *see* yet another aspect of what Lao Tzu was trying to say and gain a deeper understanding of the thoughts he was attempting to reveal.

Another reason conceptions of God are contentious is, in addition to being tied to belief, they are often entangled with religious dogma.

Many religions require followers to believe that their particular conception or definition of God represents the one and only *true* God and that all other representations or notions, or definitions of God, whether associated with other religions or not, are not only wrong but forbidden.

The power of the taboo associated with followers of any established religion even considering alternate definitions or conceptions of God is tied directly to the contours of faith, dogma and belief associated with that religion. In any religion, the statement, "our God is the one and only true God," works if, and only if, followers truly believe and have faith in that religion. In that sense, it bears no resemblance to a scientific statement of fact like "the speed of light is 186,000 miles per second" because while physicists can prove, experimentally and objectively, that the speed of light is 186,000 miles per second, it is not possible to for anyone to prove that their particular conception of God is the one and only true conception of God.

Many individuals find comfort and joy pursuing their particular faith. Some view people who do not share their beliefs as misguided, damned or infidels.

Religious experience, faith and belief in God are nondimensional experiences. That is another reason they are so contentious and why people tend to argue about them. The powerful dogma associated with religion is tied directly to the kinds of *images* we use to understand and define what our religion, faith and belief in God mean.

All the images we use to help understand who we are and how the universe works appear on our individual image screens. But, at this juncture it is important to revisit the different kinds of images we rely on with a view toward how they differ from one another and the extent to which they are attached to dimensional versus nondimensional experience. So once again I will share with you Samuel Avery's definitions of the different kinds of images we use and rely on.

> An image in time but not space is a thought; an image in time and space is a perception. An image in neither is just an image.

A thought is something that happens in time but not in space. If I think about going to the store or eating a hamburger, the images associated with those thoughts appear on my image screen and are available only to me. If I tell my wife I'm thinking about going to the store, that particular image becomes actualized for her because by telling her I've stuffed that thought into space as well as time and when the air molecules compressed by my voice impact her tympanic membranes she is able to perceive that image. Before I spoke, she could not nor could anyone else.

The cup, sitting on my desk is an image in time and space and thus, is a perception, in the same sense and for the same reasons that telling my wife I was thinking about going to the store transformed my unspoken thought/image into a perception available to her and anyone else within earshot.

For my wife, my actualized thought is an aural perceptual image. For me, the cup on my desk is a visual perceptual image.

Images in neither time nor space are more difficult to define and deal with. I associate such images with what I *see* while meditating. Occasionally, my *self* stops grabbing and lets go and the chaos spews a shard which appears as an image existing in neither time nor space. I am never able to *recognize* these images and, since I am not *myself* when they appear I cannot remember what I *saw*.

I believe the images we experience while dreaming are also neither in time nor space. My *self* participates in the dream and experiences the images associated with it. But the space and time associated with dreams is nothing like the spacetime we experience while awake. A different part of our brain is operational when we dream and acetylcholine is the primary neurotransmitter enabling the experience. Our prefrontal cortex is not participating in the dream experience so the executive functions we rely on and which define our *self* while awake are not available while we dream; except in the rare instances where dreams become lucid.

Even though dreaming is perfectly normal, while dreaming we are in a pathological state of delirium.

Images experienced during meditation and while dreaming are nondimensional. Most emotional experiences like love, hate, envy, lust, ennui and regret are nondimensional. We experience them but, not within the dimensional structure of consciousness.

And so it is with our individual experience of God. Fervent believers are convinced that their nondimensional experience of God is true and real and, for them, it is.

For atheists, their nondimensional experience of the nonexistence of God is as true and real as the experience fervent believers have.

The important point here is, both types of experience are tied directly to nondimensional images. That's the reason they can be and often are so contentious. Nondimensional images cannot be forced into the dimensions. "The Tao that can be told is not the universal Tao The name that can be named is not the universal name."

We get lost in the definitions or labels we use to understand God and our experiences of what the label God means and, in that sense, find ourselves confronted with the dilemma St. Augustine encountered when he said. "What then is time? If no one asks me, I know: if I wish to explain it to one that asketh, I know not."

R. Buckminster Fuller was fond of using the term "universal mind," which I interpreted as another word for God. He believed that when we apprehend important generalized principles operative in universe, like the lever principle, we are somehow *tuning* universal mind in much the same way radios and televisions *tune* the electromagnetic waves which carry the information they provide.

We attach labels to our concepts of God, because that's what we do. But we rarely pay attention to the images we use and assume that all the images on our image screens are essentially the same. They are not the same. Some, like thoughts, exist only in time. Others, like perceptual images, exist in time and space, and some exist in neither time nor space.

The labels, then, tend to morph and interpenetrate as we attempt to focus our attention on them in an effort to gain deeper and broader understanding. The concepts associated with the labels God, Allah, Tao, being and consciousness become entangled and since they are all, by definition, nondimensional, we cannot rely on the same tools and procedures scientists use when exploring dimensional objective phenomena which occur within the dimensional structure of consciousness.

The dimensional structure of consciousness defines and explains our dimensional experiences in the middle or classical level of reality, where we spend most of our lives. But, when we begin to focus on what happens on the quantum level of reality, even the most carefully controlled objective experiments yield results so enigmatic they appear mystical.

Objective results of experiments with entangled subatomic particles provide proof that, while the particles themselves are dimensional, in that we can measure their physical properties and manipulate them, their entanglement and the phenomena associated with their entanglement are nondimensional. They exist on the quantum level so regardless of how far they are separated in space and time, once entangled, if we disturb one particle that disturbance is instantaneously reflected in the other, even though special relativity forbids it. And, as discussed earlier, recent experiments show that entanglement applies to objects considerably larger than subatomic particles.

For those of us who enjoy thinking about what all this means we find ourselves confronted with enigmas and mystical phenomena.

Does the particle truly exist? Physicists who have explored this question tell us it exists but only if someone makes an observation. So, in a very real sense, subatomic particles are both dimensional and nondimensional. If we ignore them, we never see them. If we try to see them, we discover that we are disallowed from knowing their momentum and location simultaneously. The more we intensify our attention on them, the more nondimensional they become.

Subatomic particles appear to exist in a kind of limbo which restricts what we are allowed to know about their existence even when we trap them, accelerate them and smash them into each other. As our explorations intensify and deepen, what's truly happening on the quantum level of reality becomes unknowable.

And so it is with the nondimensional images God, Allah, Tao, consciousness and being. And here I'm reminded of another quote from *Transcendence of the Western Mind*:

> Consciousness is the only unknowable and I will call it "being" and leave it at that, hoping you know what I mean.

Since we exist in the mid or classical level of reality, when we look for some evidence of the existence of God or consciousness or being, we find no objective evidence. Our Western, Cartesian, materialist mindset prevents us from *seeing* nondimensionally.

We cling to the idea that we all *have* consciousness, but if we let go of that myth and replace it with the one which exchanges the box for the screen and posits that, instead of everyone *having* consciousness, we are all *in* consciousness, the implication becomes: consciousness is *everything*. It isn't hiding behind the rock because it is the rock. It's the blade of grass under our foot and the water that flows from our faucets. It is you and me; us and them. It is everything.

I am but one in an infinite string of beads.

Endnotes

1 Wikipedia
2 Ibid.
3 Ibid.
4 *Transcendence of the Western Mind*, Samuel Avery, 2003, Compare, p. 44
5 Ibid.
6 *The Elegant Universe*, Brian Greene, Vintage Books, March 2000 p. 50
7 *Transcendence of The Western Mind*, Samuel Avery, 2003, Compare, p. 55
8 Ibid., p. 29
9 Ibid., pp. 56-57
10 Ibid., p. 33
11 Wikipedia
12 *Transcendence of the Western Mind*, Samuel Avery, 2003, Compare, pp. 20-21
13 Ibid., p. 21
14 Ibid., p. 22
15 Ibid., p. 63
16 Ibid., p. 65
17 Ibid., p. 69
18 Ibid.
19 Ibid., p. 121
20 Google Time Line—Television adoption rate
21 *Understanding Media*, "The Extensions of Man". Marshall McLuhan, 1964,A Mentor Book
22 *Transcendence of The Western Mind*, Samuel Avery, 2003, Compare, p. 145
23 *Future Shock*, Alvin Toffler, 1970, Random House, p. 15
24 *Transcendence of the Western Mind*, Samuel Avery, 2003, Compare, p. 146
25 Ibid., pp. 147-148
26 Ibid., p. 148
27 *Tao Te Ching A new Translation & Commentary*, Ralph Allan Dale, 2002, Barnes & Noble p.23
28 Ibid., p. 29
29 *Scientific American*, June 2010, "Is Time an Illusion?" By Graig Callender, p. 59
30 *Synergetics, Explorations in the Geometry of Thinking, R. Buckminster Fuller in collaboration with E. J. Applewhite*, McMillan Publishing Co., Inc New York, 1975 pp. 440-441
31 Ibid., p. 228

32 *Transcendence of the Western Mind*, Samuel Avery, 2003 Compare, p. 63
33 *The Moral Landscape*, Sam Harris, 2010 Free Press, p. 115
34 *Living in the End Times*, Slavoj Zizek, Verso, p, 360
35 *Mind, A Brief Introduction*, John Searle, Oxford University Press, 2004, p. 65
36 Ibid.
37 From *Operating Manual for Space Ship Earth* via www.examiner.com/buckminster-fuller-in-national/operating-manual-for-spaceship-earth
38 *Buckminster Fuller's last Interview*—Part 2—The Civilization of Cyberspace Interviews... via www. Communitelligence.com/clps/clitem.cfm?adsid=173
39 *Machines Who Think*, Pamela McCorduck, W.H, Freeman and Company, San Francisco, 1972, p. 347
40 *The Information*, James Gleick, Pantheon Books, New York, 2011, p. 28
41 Ibid., p. 30
42 Ibid.
43 *Alone Together*, Sherry Turkle, Basic Books, 2011, p.203
44 Ibid., p. 168
45 http://suif.stanford.edu/-jeffop/WWW/wheeler.txt
46 *Dream Life: An Experimental Memoir*, J. Allan Hobson, The MIT Press, Cambridge Massachusetts, 2011, p. 111
47 Ibid., p 83
48 Ibid., p. 129
49 Ibid.
50 Ibid., p. 141
51 Ibid., p. 153
52 Ibid.
53 Ibid.
54 Ibid., p. 154
55 Ibid., p. 8
56 Ibid., p. 9
57 Ibid., p. 10
58 "Living in a Quantum World: Quantum mechanics is not just about teeny particles. It applies to things of all sizes: birds, plants, maybe even people." By Vlatko Vedral, *Scientific American*, June 2011.
59 Ibid.
60 Ibid.
61 Ibid.
62 Ibid.
63 Ibid.
64 Ibid.
65 Ibid.
66 Ibid.
67 Ibid.
68 *The Tao Te Ching: A New Translation & Commentary*, by Ralph Alan Dale, Verse 1 Transcending, Barnes & Noble New York, 2002

Bibliography

Avery, Samuel. *Transcendence of the Western Mind*. Compare. 2003.

Callender, Craig. "Is Time an Illusion?" *Scientific American*. June. 2010.

Dale, Ralph Allen. *The Tao Te Ching*. A new translation and Commentary. Barnes & Noble. New York. 2002.

Fuller, Buckminster R. in collaboration with E.J. Applewhite. *Synergetics, Explorations in the Geometry of Thinking*. McMillan Publishing Co., Inc. New York. 1975.

Buckminster Fuller's Last Interview. Part 2. *The Civilization of Cyberspace Interviews*. Via www.Communintelligence.com/clps/clitem.cfm?adsid=173. From *Operating Manual for Spaceship Earth*. Via www.examiner.com/buckminster-fuller-in-national/operating-manual for spaceship-earth.

Google Time Line – Television Adoption Rate.

Harris, Sam. *The Moral Landscape*. Free Press. 2010.

Hobson, Allan J. *Dream Life an Experimental Memoir*. The MIT Press. Cambridge Massachusetts. 2011.

Hofstadter, Douglas *I am a Strange Loop*. Basic Books. New York. 2007

McCorduck, Pamela. *Machines Who Think*. W.H. Freeman and Company. San Francisco. 1972.

McLuhan, Marshall. *Understanding Media: The Extensions of Man*. A Mentor Book. 1964.

Searle, John. *Mind, A Brief Introduction*. Oxford University Press. 2004.

Toffler, Alvin. *Future Shock*. Random House. 1970.

Turkle, Sherry. *Alone Together*. Basic Books. 2011.

Vedral, Vlatko. "Living in a Quantum World: Quantum Mechanics is not just about teeny particles. It applies to things of all sizes: birds, plants, maybe even people." *Scientific American*. June 2011.

Žižek, Slavoj. *Living in The End Times*. Verso. 2010.

Index